# CONTENTS

KT-436-656

## SEPARATION TECHNIQUES

### CHAPTER 1: CHROMATOGRAPHY

### CHAPTER 2: ELECTROPHORESIS

## SPECTROSCOPIC METHODS OF ANALYSIS

### CHAPTER 3: LIGHT

### CHAPTER 4: ATOMIC SPECTROSCOPY

### CHAPTER 5: VISIBLE–ULTRAVIOLET SPECTROMETRY

## CHAPTER 6: INFRARED SPECTROMETRY

## CHAPTER 7: NUCLEAR MAGNETIC RESONANCE SPECTROMETRY

## CHAPTER 8: MASS SPECTROMETRY

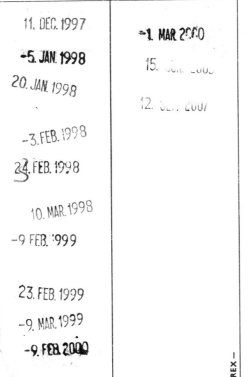
# CHEMISTRY

# De and Analysis

# E.N. Ramsden

**B.Sc., Ph.D., D.Phil.**
**Formerly of Wolfreton School, Hull**

Stanley Thornes (Publishers) Ltd.

First published 1996 by
Stanley Thornes (Publishers) Ltd,
Ellenborough House,
Wellington Street,
CHELTENHAM,
GL50 1YW

A catalogue record for this book is available from the British Library.

ISBN 0 7487 2402 8

96  97  98  99  00  /  10  9  8  7  6  5  4  3  2  1

The front cover shows the spectum of the Sun (Credit: Science Photo Library/Geoff Tompkinson).

Typeset by Tech-Set, Gateshead, Tyne & Wear
Printed and bound in Great Britain at Scotprint, Musselburgh

# PREFACE

*Detection and Analysis* has been written to match the 1996 syllabuses for the following A-level modules:

University of Cambridge Local Examinations Syndicate:
Syllabus 9254, Option 4823: Spectroscopy
Syllabus 9525, Module 4825: Methods of Analysis and Detection
University of London Examinations and Assessment Council:
Chemistry (Nuffield) Topic 18: Instrumental Methods
Oxford and Cambridge Schools Examination Board:
Syllabus 9684, Unit C8: Analytical Chemistry

Before embarking on an optional topic such as *Detection and Analysis*, students will have completed the A-level Chemistry core modules, covering atomic structure, the chemical bond and a firm foundation of physical, inorganic and organic chemistry. Should they need to revise this core material, they can consult the references to *ALC* which are to my text, *A-level Chemistry,* Third edition (Stanley Thornes). They give the section of this text in which the relevant core material can be found. Students who are using a different A-level textbook need to consult the index of their book to find the corresponding material.

## Acknowledgements

I offer my sincere thanks to Dr Rob Ritchie for his careful reading of the first draft and for his valuable comments and suggestions.

I thank the following examination boards for permission to print questions from recent A-level papers: University of Cambridge Schools Local Examinations Syndicate, University of London Examinations and Assessment Council and Northern Examinations and Assessment Board. Acknowledgement is made with each question under the abbreviations C, L and N respectively. The outline answers to the questions are my responsibility, and the examination boards accept no responsibility for them.

I thank the publishing team who have contributed their enthusiasm and expertise to the production of this book and especially the editor, John Hepburn.

My family have given me their support and encouragement all through the writing of this book.

E.N. Ramsden,
Oxford, 1996

# 1

# CHROMATOGRAPHY

## 1.1 THE PRINCIPLES

### 1.1.1 CHROMATOGRAPHY

*Chromatography is a set of techniques for separating the components of a mixture ...*

*... based on the rates at which they are carried through a stationary phase by a gaseous or liquid mobile phase.*

Chromatography was invented by a Russian botanist called Mikhail Tswett early in the twentieth century. He separated plant pigments by passing solutions of them through glass columns packed with powdered calcium carbonate or aluminium oxide. The different pigments separated into coloured bands on the column. Tswett named his technique after the Greek words *chroma* – colour – and *graphein* – to write.

Chromatography is used for the separation, identification and measurement of the chemical components in mixtures. No other method of separation is as important as chromatography. There are a variety of chromatographic techniques. All of them depend on the components of a mixture being carried at different rates through a **stationary phase** by a **mobile phase**.

### 1.1.2 ADSORPTION

*Many chromatographic techniques involve adsorption of components from the mobile phase on to the stationary phase.*

When ions or molecules in the mobile phase, e.g. a solution, approach the stationary phase, e.g. a solid surface, they may interact with atoms or molecules at the surface. The interaction is called **adsorption**. When chemical bonds are formed between a molecule or ion and the surface, the term **chemisorption**; is used. When the forces of attraction between the molecules and the surface are weaker, e.g. van der Waals forces, the term **physical adsorption** is used. The difference between adsorption and **absorption** is that in adsorption a substance is attached to the surface whereas in absorption it penetrates into the bulk. In **desorption**, an adsorbed substance becomes detached from the surface.

### 1.1.3 PARTITION

When a solute is added to a pair of immiscible liquids it may dissolve in both of them. In this case the solute will distribute itself between the two solvents. It may well be more soluble in one solvent than the other. It is found that the ratio of the two concentrations is constant:

$$\frac{[\text{concentration of solute in solvent 1}]}{[\text{concentration of solute in solvent 2}]} = k$$

*Other chromatographic techniques involve partition of solutes between two solvents.*

The constant $k$ is called the **partition coefficient** or **distribution coefficient** [See *ALC*, §8.6 for a fuller treatment of partition.]

Partition is often utilised in the preparation of organic compounds. The product is

1

partitioned between ethoxyethane (ether) and water, which are immiscible and form two layers. An aqueous solution of the organic product is shaken with ether in a separating funnel. The organic product is more soluble in ether than in water and a large fraction of it passes into the ether layer. The product can be obtained by distilling off the volatile ether. It is more efficient to use a certain volume of ether in portions for repeated extractions than to use it all at once [see *ALC*, § 8.6.2]. **Repeated extractions** with an organic solvent are effective in removing a considerable quantity of solute from an aqueous solution. This is what happens in **partition chromatography**. The technique can be applied to the separation of a number of solutes in a solution, provided that the solutes differ in their solubility in the second solvent.

*Repeated partitions take place between the mobile phase and the stationary phase, e.g. a liquid adsorbed on a solid.*

In partition chromatography many extractions are performed in succession in one operation. The solutes are partitioned between the stationary phase and the mobile phase. The stationary phase is the first solvent, e.g. water, which is adsorbed on a solid, e.g. paper or alumina gel or silica gel which has been packed into a column or spread on a glass plate. The mobile phase is a second solvent which trickles through the stationary phase.

### 1.1.4  CHROMATOGRAPHIC METHODS

#### LIQUID CHROMATOGRAPHY

The mobile phase is a liquid and the stationary phase is a liquid adsorbed on paper, solid particles packed into a column or a thin layer of porous solid. The method includes high-performance liquid chromatography and depends on either adsorption or partition.

#### GAS CHROMATOGRAPHY

The mobile phase is a gas and the stationary phase is a liquid adsorbed on a solid packed into a column. The method depends on partition.

#### ION EXCHANGE

*Methods include liquid chromatography, gas chromatography, ion exchange and gel permeation chromatography.*

Ions are separated on the basis of their ability to bond to a resin.

#### GEL PERMEATION CHROMATOGRAPHY

Substances are separated on the basis of the size of their particles, which determines their ability to penetrate the pores of a gel.

## 1.2  COLUMN CHROMATOGRAPHY

Figure 1.2A illustrates the separation of the solutes in a solution by **column chromatography**. The stationary phase is an inert solid, e.g. silica gel, alumina or cellulose or Sephadex in a glass column. Silica is slightly acidic and can adsorb basic solutes readily. Alumina is slightly basic and is a good adsorbent for acidic solutes. Sephadex is a polymer of glucose cross-linked with propane-1,2,3-triol (glycerol). The solid is made into a slurry with the solvent, poured into the column and allowed to settle. Solvents include aliphatic and aromatic hydrocarbons, alcohols, ketones, halogenocompounds, esters and mixtures of these. Any excess of solvent is run out through the tap. The stationary phase must be saturated with solvent to prevent air bubbles from forming. A solution of the **analyte** (the mixture to be analysed) is poured on to the top of the column, and the components are adsorbed at the top of the

*In column chromatography, the stationary phase is used to pack a column. A solution of the mixture is poured into the column ...*

*. . . and components are adsorbed.*

column. The mobile phase is a second solvent called the **eluant**, which carries the components of the mixture through the stationary phase. The eluant is allowed to trickle slowly through the column. The **eluate** (the solution leaving the column) is collected in fractions. If the solutes are coloured, collecting them in separate fractions is easy. If they are colourless they have to be identified in some other way, e.g. by fluorescing under UV light or by radioactive assay or by passing through a visible–ultraviolet spectrophotometer [Chapter 5] or an infrared spectrophotometer [Chapter 6] or a nuclear magnetic resonance spectrometer [Chapter 7] or a mass spectrometer [Chapter 8].

*A second solvent, the eluant, passes down the column and elutes the components in turn.*

---

**FIGURE 1.2A**
Column
Chromatography

---

*The eluate is collected in fractions and the solutes are identified by their colour, fluorescence or radioactivity, or by spectrophotometry.*

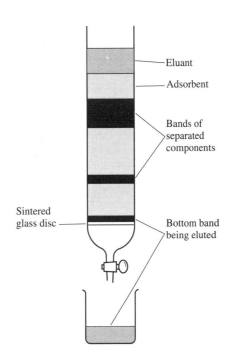

*Equilibrium is established at every level of the column between the concentration of each solute in the eluant and the concentration on the adsorbent. The separated bands are a chromatogram.*

Each solute is partitioned between the adsorbent and the eluant. The least strongly adsorbed solutes are desorbed first by the eluant and carried further down the column before being readsorbed. When fresh eluant reaches them they are desorbed again and carried further down the column. The rates at which solutes travel down the column depend on how strongly they are adsorbed. If the rates are sufficiently different the components become spread out along the column as separate bands. The separated bands are called a **chromatogram.** If the eluant is kept flowing, the solutes are **eluted** – dissolved out of the column – in turn. A complete separation may be obtained and the components in each fraction of eluate may be recovered by evaporating the solvent.

*Column chromatography is an example of liquid chromatography and adsorption chromatography.*

The technique described here is an example of **liquid chromatography** in that the mobile phase is a liquid. It is also an example of **adsorption chromatography** because the separation depends on adsorption of the solutes.

## 1.3 PAPER CHROMATOGRAPHY

*In paper chromatography a solution of the mixture is applied to a piece of chromatography paper.*

**Paper chromatography** is illustrated in Figure 1.3A. A solution of the mixture to be separated is applied to a strip of chromatography paper. This is hung in a glass tank with the end of the paper dipping into the solvent in the bottom of the tank. The solvents used include water, ethanol, butanol, glacial ethanoic acid and mixtures of these. As the solvent rises through the paper it meets the sample and the component bands spread out. The separation is stopped when the solvent has travelled nearly to the top of the paper. The distance travelled by the solvent front is measured [see Figure 1.3B]. Then for each solute the **retardation factor**, $R_F$ is calculated by

$$R_F = \frac{\text{Distance travelled by the solute}}{\text{Distance travelled by the solvent front}}$$

**FIGURE 1.3A**
Paper Chromatography

*Solvent rises through the paper and carries the components of the mixture up the paper at different rates. The distance travelled by each solute is compared with the distance travelled by the solvent front to give the $R_F$ value of the solute.*

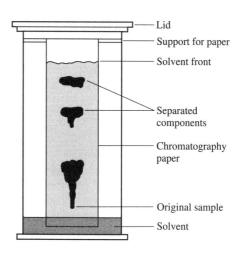

**FIGURE 1.3B**
The $R_F$ Value

*The basis of paper chromatography is partition. Solutes are partitioned between the two solvents – the solvent adsorbed on the paper and the mobile solvent.*

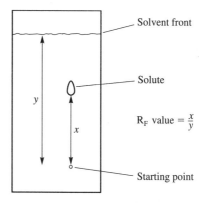

The $R_F$ values can be used to assist in identifying the components. The separated components can be obtained by cutting the paper into strips and dissolving out each compound.

What is happening in paper chromatography is based on partition. The stationary phase is the water or other solvent that is adsorbed as a film on the surface of the paper. The mobile phase is the second solvent. The solutes are partitioned between the solvent adsorbed on the paper and the mobile solvent.

*Paper chromatography can be carried out with the solvent ascending or descending . . .*

There is a design of chromatography tank which differs from Figure 1.3A in that the solvent is in a trough at the top of the tank. The paper hangs down from this trough, and solvent descends through the paper. The two versions are described as **ascending** and **descending paper chromatography**.

Two substances may have the same value of $R_F$ in a particular solvent. It is possible to separate them by drying the chromatogram and running a second chromatogram at right angles to the first in a second solvent. This is called **two-dimensional chromatography** [see Figure 1.3C].

**FIGURE 1.3C**
Two-dimensional
Chromatography

*... and can be run in two
directions with two
solvents ...
... as two-dimensional
paper chromatography.*

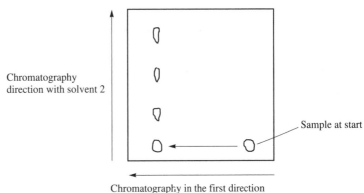

Chromatography
direction with solvent 2

Sample at start

Chromatography in the first direction
with solvent 1 failed to separate the mixture.

The uses of paper chromatography include the separation of pigments in inks and
food colourings and dyes.

## 1.4 THIN LAYER CHROMATOGRAPHY

*In thin layer
chromatography the solid
adsorbent is spread as a
thin layer on a glass plate.*

Another version of liquid chromatography is **thin layer chromatography (TLC)**. The
solid adsorbent is in the form of a thin layer on the surface of a glass plate. The
adsorbent, e.g. silica gel or calcium sulphate, is made into a thick paste with water and
spread evenly over a glass plate. The thin layer of paste is allowed to dry and baked in
an oven. Spots of the mixture are applied and a chromatogram is developed in the
same manner as for paper chromatography. Thin layer chromatography has the
advantage that a variety of adsorbents can be used.

**FIGURE 1.4A**
Thin Layer
Chromatography

*TLC is carried out in a
similar way to paper
chromatography ...
... and it can be run in two
dimensions*

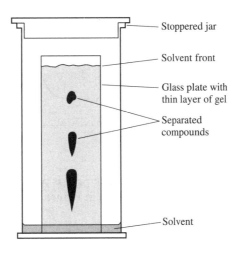

Stoppered jar

Solvent front

Glass plate with
thin layer of gel

Separated
compounds

Solvent

*TLC has advantages ...
... a variety of absorbents
can be used ...
... and the separations are
more efficient.*

The particle size of the stationary phase is smaller in thin layer chromatography than
in paper chromatography. As a result the separations are much more efficient and
more reproducible. Often separations can be achieved in a few centimetres, and coated
microscope slides are frequently used for TLC.

*TLC is widely used in the
pharmaceutical industry,
e.g. for amino acid
analysis.*

Thin layer chromatography is widely used in the drug industry for the determination
of product purity. TLC is used in clinical laboratories and biochemical laboratories. It
can be used to follow the progress of a reaction. The number of analyses performed
by thin layer chromatography is at least as great as by high-performance liquid

chromatography [§ 1.7]. TLC is used for the separation and identification of amino acids; It can be run in two dimensions, as for paper chromatography [§ 1.3 and Figure 1.4B].

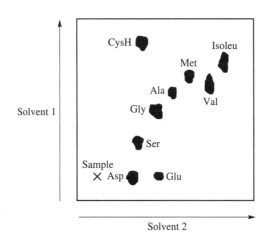

Many spots are not visible unless the plates are 'developed'. This is done by spraying with a solution which reacts with the solutes to make them visible. A solution of iodine is used to reveal the presence of some aromatic compounds with electron-donating groups, e.g. $C_6H_5NH_2$. Ninhydrin is used to identify amino acids. Some spots become visible under a UV lamp. Radioactive solutes can be identified on a TLC plate by passing the plate under a Geiger counter. A recorder plots the count rate as the plate passes under the counter [Figure 1.4C].

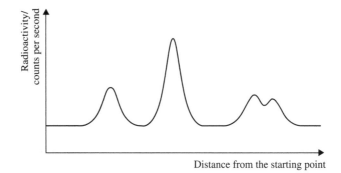

The area under each peak is proportional to the amount of that solute. It can be measured as shown in Figure 1.4D

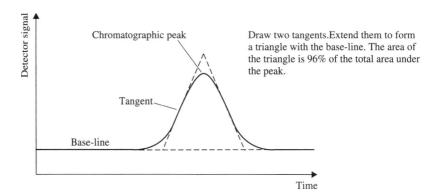

**1.** Explain the terms (*a*) mobile phase, (*b*) stationary phase, (*c*) eluant, (*d*) liquid chromatography, (*e*) partition (*f*) retardation factor, (*g*) adsorption.

**2.** Explain (*a*) what is meant by two-dimensional chromatography, (*b*) the difference between ascending and descending paper chromatography.

**3.** (*a*) Suggest two explanations for the single spot in this thin layer chromatogram.
(*b*) Say how you would find out which explanation is correct.

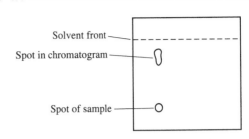

Solvent front
Spot in chromatogram
Spot of sample

# 1.5 GAS CHROMATOGRAPHY

*In gas chromatography the mobile phase is a gas. The liquid which forms the stationary phase is spread on the surface of solid particles . . .*

In **gas chromatography** (**GC**) the mobile phase is a gas (or a volatile liquid or solid). It is separated into components by equilibration with a liquid, which is the stationary phase. The liquid is spread on the surface of inert solid particles which pack a long (5–10 m) narrow (2–10 mm bore) column. The column is coiled so that it will fit into a thermostatically controlled oven. The sample is injected through a self-sealing disc into a heated chamber where it vaporises. The injection chamber is 50–100 °C above the temperature of the column. An inert gas – nitrogen or a noble gas – carries the sample through the column. Each component sets up a partition equilibrium between the vapour phase and the liquid phase. Some components are less soluble in the liquid than others and emerge from the column ahead of the others. A detector records each component as it leaves the column.

---
**FIGURE 1.5A**
Gas Chromatography
---

*. . . which are packed into a column.
Each component is partitioned between the vapour phase and the liquid phase.*

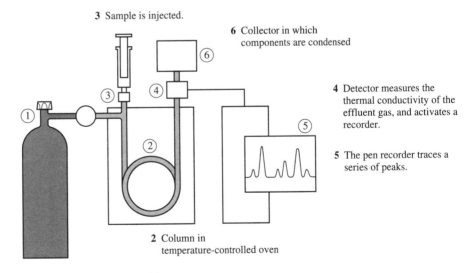

**3** Sample is injected.

**6** Collector in which components are condensed

**4** Detector measures the thermal conductivity of the effluent gas, and activates a recorder.

**5** The pen recorder traces a series of peaks.

**2** Column in temperature-controlled oven

**1** Cylinder of carrier gas with valve to control the flow rate

## 1.5.1 DETECTORS FOR GAS CHROMATOGRAPHY

Detection devices must respond to minute concentrations of substances as they leave the column. The solute concentration in the carrier gas is no greater than a few parts per thousand. A peak may pass a detector in one second or less so the detector must respond rapidly. In addition, a linear response (with a signal proportional to the concentration of substance) is desirable.

The position of each peak in a chromatogram is determined by the **retention time** of the component. The retention time is the time between the start of the chromatographic separation and the appearance of the peak at the detector [Figure 1.5B]. The retention time depends on the solvent and the temperature. In order to use the retention time to identify a component, it is necessary to compare its retention time with that of a known sample of the substance under identical conditions. **Dead time** is the time for a species which is not adsorbed to pass through the column or cell or plate.

**FIGURE 1.5B**
Retention Time

*A detector monitors the components as they leave the column.*

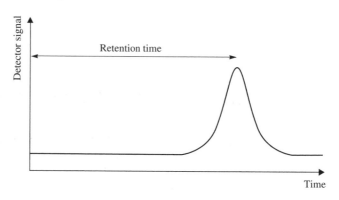

## 1.5.2  THERMAL CONDUCTIVITY DETECTOR

The thermal conductivity of the carrier gas stream changes when an eluate is present. The sensing device is an electrically heated element of tungsten wire. The resistance of the wire depends on its temperature, which is governed by the rate at which the surrounding gas conducts heat away from the element to the walls of the metal block in which it is housed [Figure 1.5C].

**FIGURE 1.5C**
A Thermal Conductivity Detector

*A thermal conductivity detector works on the principle that the resistance of a wire depends on its temperature which depends on the thermal conductivity of the surrounding gas.*

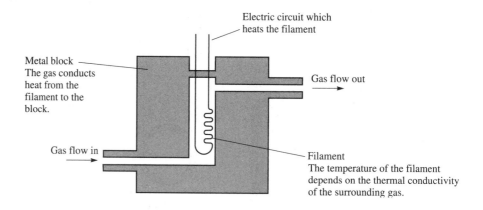

Thermal conductivity detectors are simple and respond to a large range of organic and inorganic compounds. They have the advantage that they do not destroy the samples. The sensitivity is lower than for other detectors: about $10^{-8}$ g solute per cm$^3$ of carrier gas.

## 1.5.3  FLAME IONISATION DETECTOR

Most organic compounds when pyrolysed at the temperature of a hydrogen–air flame produce ions and electrons that conduct electricity. In the **flame ionisation detector** shown in Figure 1.5D, the ions are attracted to and captured by a detector. The current that results is amplified and recorded. The detector is insensitive to non-combustible gases, e.g. $H_2O$, $SO_2$, $CO_2$, $NO_2$. The flame ionisation detector is a useful detector for most organic samples. It is the most widely used detector because of its high sensitivity ($\sim 10^{-13}\,\mathrm{g\,m^{-3}}$) which means that it can be used in capillary columns which do not supply enough sample for a thermal conductivity detector. It has a linear response (current proportional to concentration) over a range of $10^7$ in concentration. The flame ionisation detector has the disadvantage that it destroys the sample.

**FIGURE 1.5D**
A Flame Ionisation
Detector

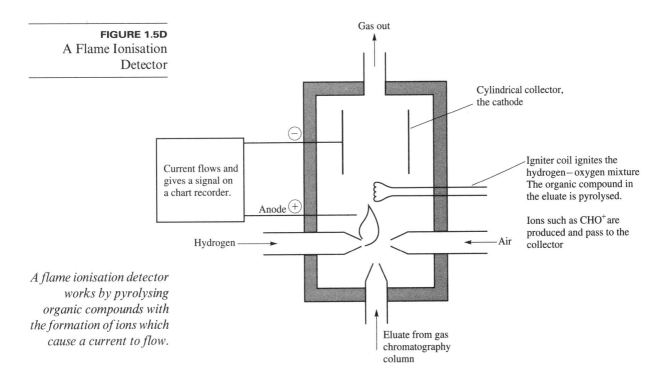

*A flame ionisation detector works by pyrolysing organic compounds with the formation of ions which cause a current to flow.*

### 1.5.4  ELECTRON CAPTURE DETECTORS

*In electron capture detectors the eluate is passed through a beam of electrons which flows between two electrodes. Organic compounds capture electrons and decrease the current.*

In an **electron capture detector** the eluate from a column is passed over a $\beta$-emitter. An electron from the emitter causes ionisation of the carrier gas and produces a burst of electrons. This maintains a constant current between a pair of electrodes. The current decreases in the presence of organic compounds, which capture electrons.

The electron capture detector is selective: it is sensitive to electronegative functional groups, e.g. halogen, and insensitive to compounds such as hydrocarbons, amines and alcohols. It is a useful method for the detection and determination of pesticides that contain chlorine. Electron capture detectors are very sensitive and do not consume the sample as flame detectors do.

### 1.5.5  SELECTIVE DETECTORS

*Elution is a process in which solutes are washed through a stationary phase by the movement of a mobile phase.*

Gas chromatography is often coupled with mass spectrometry, nuclear magnetic resonance and infrared spectrometry. These 'hyphenated methods', e.g. GC-MS and GC-IR, are powerful methods of identifying the components of complex mixtures. When these methods were first used, fractions of eluate were collected and then investigated by IR, MS or NMR. Now spectrometers are used to monitor the eluate

*Gas chromatography is often coupled with spectrometers: mass spectrometer, NMR spectrometer and IR spectrometer ...*

continuously. The instruments are computerised with a large computer memory in which are stored spectral data for comparison with and identification of the spectrum of the compounds in the eluate.

Applications of gas chromatography include analysis of vehicle exhaust gases, pesticides, sedative mixtures and steroids. Lipids are analysed by hydrolysing them to fatty acids, esterifying the fatty acids with methanol and performing chromatography on the methyl esters [see Figures 1.5E and F].

**FIGURE 1.5E**
A Chromatogram of
Exhaust Gases

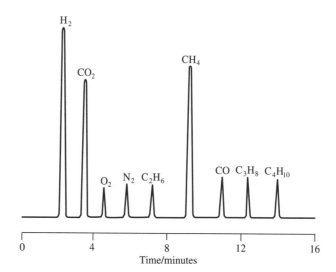

*... which can monitor an eluate continuously and identify a component by comparing its spectrum with a computer's database.*

**FIGURE 1.5F**
A Chromatogram of
Methyl Esters of Fatty
Acids (A = caprylate,
B = laurate,
C = myristate,
D = palmitate,
E = stearate, F = oleate)

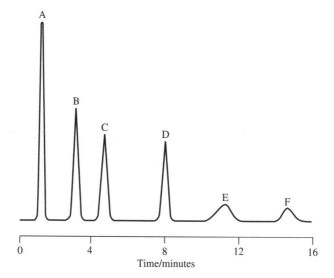

## 1.5.6  FORENSIC CHEMISTRY

In November 1974, bombs exploded in two pubs in Birmingham, killing 21 people. Three days later the police arrested six Irishmen who were on their way to attend the funeral of an IRA man who had blown himself up with his own bomb. The 'Birmingham Six' were convicted of the murders and sent to prison. The prosecution placed strong emphasis on the results of forensic tests of swabs taken from the suspects' hands. The Home Office forensic scientist, Dr Skuse, used a spot test, called the Geiss test, to test for nitroglycerine, a component of explosives. When the Geiss reagent is added, nitroglycerine gives a pink colour. He followed this with thin layer chromatography and with gas chromatography–mass spectrometry. With TLC he obtained negative results. With GC-MS he obtained positive results with two of the suspects. Dr Skuse was never able to produce the printout of the GC-MS in court because 'it was missing'. Dr Black, a Home Office explosives expert, claimed that other substances, such as nitrocellulose, would give a positive result in the Geiss test, and that nitrocellulose could be picked up from lacquers and varnishes. His evidence was rejected, and the certainty of Dr Skuse's opinion that two of the six had handled explosives led to the conviction of the six.

*The conviction of the 'Birmingham Six' was based on flawed chromatographic evidence.*

Five years later, research showed that a wide range of substances could give a positive result in the Geiss test, including substances used to coat playing cards and cigarette packets. A TV documentary featured this research and was followed by the Home Secretary granting the Birmingham Six leave to appeal. The appeal was dismissed.

In 1991 a second appeal was heard. It came out that back in 1974 forensic scientists other than Dr Skuse had used the Geiss test on some of the six suspects and obtained positive results which were tracked down to a substance in adhesive tape. The reason why these Home Office forensic scientists had never provided these facts about adhesive tape to the defence between 1974 and 1990 was never explained. Respect for the forensic expert and his flawed analytical work seems to have paralysed everyone. In 1991 the conviction was overturned and the Birmingham Six were set free.

The message seems to be that when working in the laboratory you should always keep an on-the-spot record of your work and you should keep your results in a well-organised filing system. Powerful analytical tools will not do the job without an efficient operator. There is also a message about the temptation to fudge results. Six Irishmen on their way to an IRA bomber's funeral were prime suspects. Investigators assumed they were guilty and did not scrutinise the evidence as honestly as they should have done.

## 1.6  HIGH-PERFORMANCE LIQUID CHROMATOGRAPHY

*The efficiency of liquid chromatography is increased by decreasing the size of the solid particles on which the liquid is adsorbed . . .*

Liquid chromatography [§ 1.2] is carried out in glass columns with diameters of 1–5 cm and length 50–500 cm. The diameters of the particles are 150–200 mm, and flow rates are a fraction of one cubic centimetre per minute. Separation times are long, often some hours. Increasing the rate of flow by pumping is not effective because it decreases the efficiency of the column. However a decrease in particle size increases the efficiency of the column. A technique for using particles of diameter as small as 10 $\mu$m has been developed. The efficiency of a column packed with these tiny particles is so high that pumping pressures of several hundred atmospheres can be used to increase flow rates. The technique is called **high-performance liquid chromatography (HPLC)**.

*. . . in high-performance liquid chromatography.*

*The eluate is analysed by visible–ultraviolet spectrometry, or mass spectrometry, IR spectrometry or by some other measurement of a physical quantity.*

The columns are usually made of stainless steel tubing, of length 10–30 cm and inside diameter 5–10 mm and packed with small particles . The commonest packing is silica particles. Other packing materials are alumina, porous polymer particles and ion exchange resins. Pressures of up to 100 atm are used. A sample of 5–20 mm$^3$ is injected into the top of the column, and the solutes are identified by spectroscopy. The most widely used detectors in HPLC are based on absorption of UV or visible radiation [see Figure 1.6A]. Other detectors use mass spectrometry, infrared spectrometry, fluorescence, conductivity on refractive index measurement.

**FIGURE 1.6A**
A UV Detector for HPLC

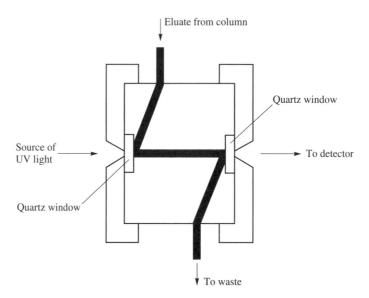

## 1.6.1   COMPARISON OF GC AND HPLC

Both gas chromatography and high-performance liquid chromatography:

- are efficient and widely applicable
- require only a small quantity of sample
- can be readily adapted to quantitative analysis
- can be interfaced with mass spectrometry and other detectors
- can be used to identify solutes from their retention times under fixed conditions.

*The advantages of GC and*   Advantages of HPLC are:
*HPLC are compared.*

- It is run at a lower temperature than GC and can be used for compounds which decompose at the temperatures needed for GC.

- It can be used for inorganic ions.

Advantages of GC are:

- The equipment is simple and inexpensive.
- It is rapid.
- Resolution is very good.

## 1.6.2   ANALYSIS BY HPLC OF DECAFFEINATED COFFEE AND CHOCOLATE PRODUCTS

Caffeine occurs in tea (2–3.5% of the mass of dry leaves) and coffee (1.1–1.8% in roasted coffee beans; 3% in instant coffee). Some people are worried about the stimulant and diuretic effects of caffeine and prefer decaffeinated products. Caffeine is dissolved out of coffee in liquid carbon dioxide. It is important to be able to find the content of caffeine before and after processing. The food industry uses chromatography to check the quality of caffeine-containing products and chocolate products.

Caffeine

Theobromine

Theobromine is formed from caffeine by metabolic processes in the body. It occurs naturally in cacao beans which are used to make cocoa and chocolate. The theobromine content of defatted cocoa powder is 2.5% by mass. It is possible to calculate the content of cocoa solids in chocolate products from the theobromine content.

*HPLC linked to a UV spectrophotometer is used to analyse coffee, tea and chocolate for their content of caffeine and theobromine.*

Caffeine and theobromine can be determined together. A known mass of coffee, tea or chocolate product is boiled with water, clarified and made up to a known volume. A portion of the solution is analysed on an HPLC column. The eluate can be passed through a UV detector. Both caffeine and theobromine absorb at 274 nm. The concentrations of caffeine and theobromine are found by comparing their retention times with those of caffeine and theobromine in standard solutions.

## 1.6.3 DETECTION OF POLLUTANTS BY GC AND HPLC

*GC and HPLC are used for the detection and analysis of air pollutants and water pollutants.*

Gas chromatography is used for the detection and analysis of air pollutants, e.g. sulphur dioxide and carbon monoxide.

It is also used for the detection and analysis of organic compounds which are water pollutants. In conjunction with mass spectrometry it is used to detect herbicides, e.g. 2,4-dichlorophenoxyethanoic acid. GC cannot be used to analyse organic compounds in water directly because the concentration is too low. The organic compounds are removed by passing a purging gas, e.g. helium, through a sample of the water and concentrating the purged gases on a column before driving them off by heating into the GC column.

HPLC has the advantage in that the materials to be analysed need not be in the vapour phase and it has become a useful technique for the analysis of water pollutants.

## 1.6.4 CHROMATOGRAPHY AT THE RACE COURSE

*Methods of testing for drugs include column chromatography, TLC, GC and HPLC.*

The rules of horseracing require horses to be tested for illegal drugs. These may be drugs to fight an infection, anabolic steroids to build up muscles, painkillers to allow an unfit horse to run, stimulants to improve performance and sedatives to impair performance. It is usual to test winners and also any horse which runs poorly, for instance a favourite which finishes poorly and may have been doped with sedatives. The horse's urine is analysed by the Horseracing Forensic Laboratory in Newmarket. Similar procedures are used for human athletes.

There are so many components in horse urine that four different types of chromatography are employed.

● Column chromatography is used to separate the horse urine into four fractions: one strongly acidic, one weakly acidic, one basic and one neutral. Three more types of chromatography are employed to separate the compounds in these fractions.

● TLC is used for the strongly acidic fraction. A standard is run containing all the normal components of urine that should show up in the chromatogram. Any spots other than those in the standard are investigated.

● HPLC is used for the weakly acidic fraction and the neutral fraction. The eluates are passed through UV microcells. The wavelength of each peak and the retention time [see § 1.5.1] are recorded and the results are fed into a computer which can compare the information with the characteristics of about 40 drugs.

● GC is used for the basic fraction. The bases in the eluate are identified by their retention times and their mass spectra.

1. (*a*) What is a fatty acid?

(*b*) An unknown lipid is hydrolysed to glycerol and a mixture of fatty acids. How could you determine the composition of the mixture of fatty acids?

2. (*a*) Explain how thermal conductivity detectors and flame ionisation detectors work.

(*b*) What is the advantage of each type of detector?

(*c*) What other types of detector are used with gas chromatography?

(*d*) Explain the terms (i) retention time and (ii) dead time.

3. (*a*) Why is HPLC superior to other types of liquid chromatography?

(*b*) How does the construction of HPLC columns differ from that of other liquid chromatography columns?

(*c*) For what types of compounds is HPLC preferred to GC?

(*d*) What advantages does GC have over HPLC?

## 1.7  ION EXCHANGE CHROMATOGRAPHY

*Ion exchange chromatography removes ions of certain types from solution and replaces them with different ions by means of anion exchange resins.*

**Ion exchange chromatography** is used to remove ions of one type from a mixture and replace them with different ions. A chromatographic column is filled with granules of a synthetic resin which contains charged groups, either positively charged groups (in anion exchange resins) or negatively charged groups (in cation exchange resins). Many resins are based on polystyrene with cross-linking. Some resins contain quaternary ammonium groups, e.g. $-\overset{+}{N}(CH_3)_3\ OH^-$; these resins are strongly basic and can remove anions, e.g. $NO_3^-$, $I^-$, $Br^-$, and replace them with hydroxide ions. Some resins contain the strongly acidic sulphonic acid group, $-SO_3^-H^+$, while others contain the weakly acidic carboxyl group, $-CO_2H$. These resins can exchange cations for hydrogen ions. Some applications of ion exchange chromatography are described below.

### 1.7.1  WATER SOFTENERS

In ion exchange an equilibrium is set up. Two ions compete for the binding sites on a resin. For example the **water softener** Permutit® is sodium aluminium silicate, an ion exchange resin which can replace calcium ions in hard water by sodium ions.

*Water softeners remove calcium ions and magnesium ions and replace them with sodium ions.*

$$2Na^+\,(resin) + Ca^{2+}\,(aq) \rightleftharpoons Ca^{2+}\,(resin) + 2Na^+\,(aq)$$

An equilibrium constant $K$ can be written:

$$K = \frac{[Ca^{2+}\,(resin)]\,[Na^+\,(aq)]^2}{[Na^+\,(resin)]^2\,[Ca^{2+}\,(aq)]}$$

The equilibrium position is different at each level of the ion exchange column. At the top of the column, the concentration of calcium ions in solution is high, and the equilibrium position moves to the right-hand side of the equation: calcium ions are adsorbed. Eventually the ability of the column to supply sodium ions is exhausted. The column must be regenerated. If a concentrated solution of sodium chloride is now passed down the column, the position of equilibrium moves towards the left-hand side: calcium ions are desorbed and sodium ions are adsorbed to regenerate the column.

### 1.7.2  PURIFICATION OF WATER

*Deionised water is obtained by passing water through an anion exchanger and a cation exchanger.*

Water can be passed through a **cation exchanger**, containing e.g. $-SO_3^-H^+$ groups, to replace cations by hydrogen ions and through an **anion exchanger**, containing e.g. $-\overset{+}{N}(CH_3)_3\,OH^-$, to replace anions by hydroxide ions. If tap water is run slowly through an ions exchange resin, **deionised water** of high purity can be obtained.

### 1.7.3  VOLUMETRIC ANALYSIS

*Ion exchange can be applied to volumetric analysis.*

The concentration of ions in a solution can be found. For example, in a saturated solution of calcium sulphate, the calcium ion can be adsorbed and replaced with hydrogen ions which can be found by titration against a standard alkali.

### 1.7.4  ANALYSIS OF A MIXTURE OF AMINO ACIDS

*A mixture of amino acids can be analysed by absorbing them on an anionic resin at low pH and eluting them in turn.*

To analyse a mixture of amino acids, a cation exchange column filled with sulphonated poly(styrene) resin may be used. If the separation is carried out at pH 3, the sulphonic acid groups in the resin are ionised as $-SO_3^-H^+$, and most of the amino acids are cations, e.g. $H_3\overset{+}{N}CH_2CO_2H$. The most basic amino acids, e.g. lysine, histidine, are bound most tightly, and the most acidic amino acids, e.g. glutamic acid, aspartic acid, are bound least tightly. When an eluant, e.g. aqueous sodium chloride, is passed down the column, some amino acids are displaced more readily than others and are the first to flow out of the column in the eluate. The procedure has been automated in an apparatus called an **amino acid analyser**. It enables the amounts of all the amino acids present in a protein to be found in a few hours, using less than a milligram of protein.

### 1.7.5  THE DISHWASHER

*An automatic dishwasher requires a water softener...*

*... and a cationic exchange resin is used to remove calcium and magnesium ions.*

Automatic dishwashers incorporate water softeners. When the water in a dishwasher is heated, calcium and magnesium hydrogencarbonates form carbonates which are deposited on the heating element, reducing its efficiency. They can also deposit a film on the crockery in the dishwasher. To remove calcium and magnesium ions from the water entering the dishwasher, a cationic exchange resin is used to exchange these ions for sodium ions [see § 1.7.1]. The resin is regenerated by passing a saturated solution of sodium chloride through it and pumping out the displaced calcium ions and excess sodium chloride. The solution of sodium chloride is made by pumping water through a container full of sodium chloride, which must be refilled from time to time. Table salt contains magnesium chloride, so for dishwashers it is better to use granular salt or dendritic salt.

# 1.8  GEL PERMEATION CHROMATOGRAPHY

Another type of column chromatography is **gel permeation chromatography**, or **gel filtration**. It differs from other types of chromatography in that no equilibrium is set up between solute and stationary phase. The method is used to separate substances with large molecules. The mixture passes as liquid or gas through a porous gel. Figure 1.8A shows the separation of proteins on a porous gel which consists of particles that are a network of polysaccharide fibres. A solution of proteins is applied to the column. The largest protein molecules cannot penetrate the pores in the particles and pass quickly down the column. Smaller protein molecules can enter some of the larger pores. The smallest protein molecules can penetrate the smallest pores in the particles. The mixture of solutes is washed from the column by an eluant. Separation occurs according to molecular size [see Figure 1.8A].

*Gel permeation chromatography or gel filtration is a type of chromatography in which no equilibrium is set up.*

**FIGURE 1.8A**
Gel Permeation
Chromatography

*Substances with large molecules are separated on the basis of the size of their molecules.*

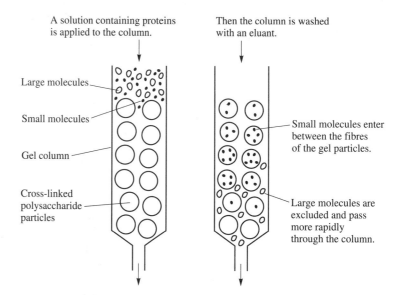

Calcium phosphate gel, alumina gel, starch and hydroxyapatite are used as adsorbents. Different gels are available which allow the separation of proteins with molar masses ranging from a few hundred to over a hundred million. The greatest separation is achieved by using very small gel particles, but the rate of flow through the column is then very slow.

## 1.8.1  USING GEL PERMEATION CHROMATOGRAPHY TO TEST SOAPS AND CREAMS

*The technique is used to find out whether substances are photoallergens.*

Some substances are irritants; they cause irritation of the skin, e.g. itching, reddening and cracking. Some substances are **allergens**; they do not irritate everyone, only those individuals who show an **allergic reaction** to the substance. This allergic reaction is a **learned response**: it does not happen with the first exposure to the allergen. In some cases the allergens cause their effects by means of photochemical reactions; these are called **photoallergens**. They absorb light from sunlight to form a species which bonds to protein molecules in the skin. Many of the photoallergens which have been identified were found as the result of research on soaps and medical formulations that had been found to cause allergic reactions. These photoallergens had been added by manufacturers on account of their germicidal action. Now that the action of photoallergens is better understood, new chemicals are screened before addition to products. This has been done by testing on animals, but an *in vitro* (in glass) method can be used.

Formula of a germicide which
was added to toilet soaps until
it was found to be photoallergenic

*The test substance is mixed with the protein human serum albumin ...* The protein human serum albumin is present in skin and blood and can be extracted from blood. It is buffered and added to a solution of the test chemical in a quartz cell. The UV spectra [see Chapter 5] of the protein, the test substance and the mixture of the two are taken. Then the mixture is irradiated for 30 minutes with UV light of a wavelength that is absorbed by the test substance. To find out whether bonding has occurred, the protein fraction is separated by gel permeation chromatography and the UV spectrum of the protein is obtained. If the spectrum of *... the protein is separated by gel permeation chromatography and its UV spectrum is obtained to find out whether the protein has changed through exposure to the test substance.* the protein is different from human serum albumin, there may be bonding of the test chemical to the protein or some other interaction. If different separation techniques, with different solvents, always give a spectrum which is different from human serum albumin alone, it indicates that the test chemical has bonded covalently to the protein. The test substance is then rejected as an ingredient of any products for use on the skin. This *in vitro* testing of possible photoallergens is faster and cheaper than testing on animals and satisifies people who object to testing products on animals.

## CHECKPOINT 1.8

**1.** The figure shows a chromatogram obtained by separating ions on a cation exchange resin, using an electrical conductivity detector.

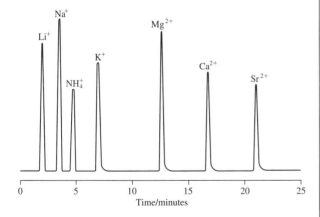

In a second sample run with the same solvent under identical conditions peaks were observed with retention times 7 minutes and 21 minutes. Which cations were present?

**2.** (*a*) Why is ion exchange described as an equilibrium?

(*b*) Give an example of (i) a group which is a cation exchanger and (ii) a group which is an anion exchanger. Write equations for the equilibria involved.

**3.** Explain how a water softener works.

**4.** Explain how a cation exchanger could be used to find the concentration of $Ca^{2+}$ and $Mg^{2+}$ in hard water.

**5.** Explain why ion exchange can be used for the separation of amino acids.

**6.** The figure shows a plot of the concentration of protein against the volume of eluant passed through a gel filtration column.

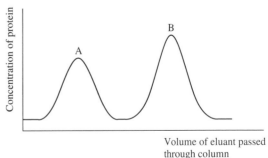

Which of the peaks represents the largest molecules?

**7.** Why can proteins be separated by gel permeation chromatography whereas amino acids cannot?

## 1.9  AFFINITY CHROMATOGRAPHY

Separating macromolecular substances is an important part of the biochemist's work. The methods used are based on differences in solubility and in the charge and size of particles. Sometimes a method does not give results because the components of the mixture have one of these properties in common. Often a combination of techniques is needed because it is unlikely that the components will have the same combination of properties – charge, size and solubility.

A different approach is to utilise a specific binding site in the component which it is required to separate. For example, an enzyme has a specific binding site for its substrate. In a mixture of enzymes extracted from a tissue, it is likely that a certain substrate will bind to one enzyme only or to a group of closely related enzymes. This is the basis of **affinity chromatography**. For substances other than enzymes a ligand which binds specifically to the required substance is used [see Figure 1.9A].

**FIGURE 1.9A**
Affinity Chromatography

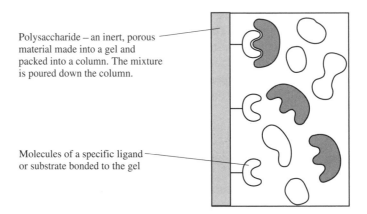

Polysaccharide – an inert, porous material made into a gel and packed into a column. The mixture is poured down the column.

Molecules of a specific ligand or substrate bonded to the gel

*Affinity chromatography uses a column with binding sites that are specific for the component which is required to be separated from a mixture. Other substances do not bond and pass through the column. Later the desired substance is eluted.*

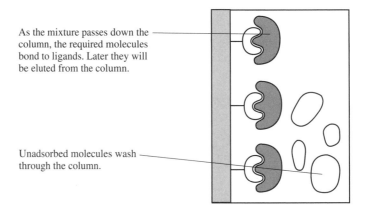

As the mixture passes down the column, the required molecules bond to ligands. Later they will be eluted from the column.

Unadsorbed molecules wash through the column.

In practice difficulties are encountered with affinity chromatography. It may be difficult to bond the ligand to the gel. Ligands may bind other components in addition to the required component. Some substances may be adsorbed on the gel. When there is a very specific relationship between ligand and component it can be a very effective method. It can also be of value in combination with other separation methods.

## QUESTIONS ON CHAPTER 1

**1.** 'Partition of a solute between two solvents is the basis of paper chromatography.'
Explain this statement.

**2.** 'The retardation factor of pigment A, using methanol–ethanoic acid as eluant, is 0.65.'

(*a*) Explain this statement.

(*b*) Explain how you would try to separate A from pigment B which has the same value of $R_F$.

**3.** A mixture of amino acids gives the chromatogram shown in the figure. How could you check to find out whether there were no more than four different amino acids in the mixture?

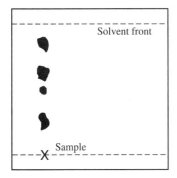

**4.** (*a*) Suggest three features of thin layer chromatography that make it a valuable technique.

(*b*) How can TLC be used (i) to identify the components of a mixture and (ii) to obtain the separated components?

**5.** (*a*) Explain the part played by partition equilibria in gas chromatography.

(*b*) Detectors used in gas chromatography include (i) thermal conductivity detectors, (ii) flame ionisation detectors and (iii) electron capture detectors. State one advantage and one drawback of each method.

**6.** An athlete is suspected of taking illegal steroids. Outline how in principle gas chromatography could be used to find out the truth of the matter.

**7.** (*a*) What is the advantage of high-performance liquid chromatography?

(*b*) What design feature is responsible?

**8.** (*a*) Ion exchange chromatography depends on setting up an equilibrium. Give two examples, with equations.

(*b*) Explain how deionised water is produced.

(*c*) Explain how an amino acid analyser works.

**9.** An anion exchange chromatogram is shown in the figure.

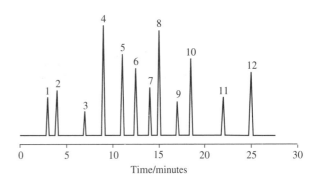

1 = fluoride, 2 = ethanoate, 3 = methanoate,
4 = chloroethanoate, 5 = chloride, 6 = nitrite,
7 = dichloroethanoate, 8 = hydrogenphosphate, $HPO_3^{2-}$,
9 = nitrate, 10 = sulphate, 11 = phosphate, $PO_4^{3-}$,
12 = citrate.

An unknown sample run under the same conditions showed peaks with retention times 9 minutes, 17 minutes and 25 minutes. Identify the components responsible for these peaks.

**10.** An elution profile from a gel filtration experiment on a mixture of proteins is shown in the figure.

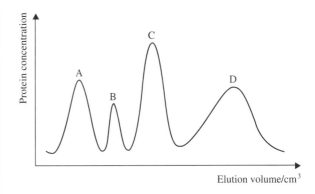

(*a*) How many proteins does the mixture contain?

(*b*) Rate the proteins in order of increasing molar mass.

(*c*) How could you check to find out whether peak C contains a single protein or a mixture?

# 2

# ELECTROPHORESIS

## 2.1  SEPARATION ON THE BASIS OF CHARGE

*Colloidal particles migrate under the influence of an electric field . . .*
*. . . a movement called electrophoresis . . .*

Colloids are heterogeneous mixtures whose particles are larger than the molecules and ions that form solutions and smaller than the particles which form suspensions [see *ALC*, §9.5]. Many biologically important macromolecular substances are colloidal. Under the influence of an electric field colloidal particles migrate. They are attracted to one electrode and repelled by the other. This movement under the influence of an electric field is called **electrophoresis**. Since colloidal particles of different types migrate at different rates, electrophoresis can be used to separate different substances [see Figure 2.1A].

*. . . which can be used to separate substances that migrate at different rates.*

The mixture is applied to a support material, which may be absorbent paper or a gel made from starch or a synthetic polymer. The gel is saturated with solvent. When an electric field is applied, charged particles move towards the electrode of opposite charge at a speed which depends on their charge. When separation has occurred, the paper or gel is removed from the apparatus. The components are located by means of staining or in some cases by viewing under UV light. The gel or paper is cut up into sections containing individual components, and these are dissolved out.

*Separation takes place on paper or on a gel to which an electric field is applied.*

Electrophoresis is the most widely used technique for separating macromolecules on the basis of charge. It separates small quantities and is used for analysis. It is not usually suitable for the production of large quantities of material.

### 2.1.1  EFFECT OF pH

*Charged molecules move towards the electrode of opposite charge at speeds which are proportional to the charges on the molecules.*

Electrophoresis is used for separating proteins. Protein molecules contain basic and acidic groups. At low pH, basic groups, e.g. $-NH_2$, are ionised:

$$-NH_2\,(aq) + H^+(aq) \rightleftharpoons -NH_3{}^+\,(aq)$$

At high pH, acid groups, e.g. carboxyl groups, are ionised:

$$-CO_2H\,(aq) \rightleftharpoons -CO_2{}^-\,(aq) + H^+\,(aq)$$

*The charge on a protein molecule and its behaviour in electrophoresis depend on the pH of the solution.*

The overall charge on a protein molecule therefore depends on the pH of the medium. At low pH a certain protein may move towards the negative electrode, whereas at high pH it may move towards the positive electrode. The speed at which the protein moves depends on the exact value of the pH. A group of proteins which move in the same direction at the same speed at one pH may be separated by a second electrophoresis at a different pH.

20

**FIGURE 2.1A**
Electrophoresis of a
Mixture of Proteins, e.g.
Blood

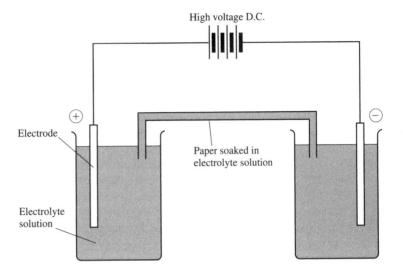

(a) Apparatus for electrophoresis

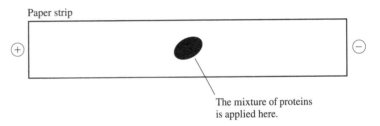

(b) Applying a liquid containing proteins, e.g. blood, to the paper strip

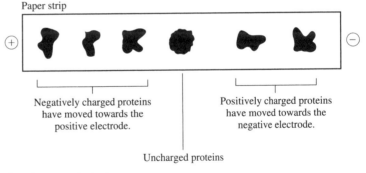

(c) Electrophoresis pattern obtained

## 2.2   SEPARATION ON THE BASIS OF SIZE

A modification of electrophoresis is **gel electrophoresis**. It is widely used in molecular biology. This type of electrophoresis separates substances on the basis of the size of their molecules. Figure 2.2A shows an apparatus which can be used for separating macromolecules such as DNA molecules of different sizes. These molecules are all negatively charged. The method requires a thin slab of a polysaccharide gel with wells (small slots) into which samples can be placed. This is prepared by allowing the liquid gel to solidify with a mould of the required shape on top. Samples are placed in the wells. An electric field is applied and DNA molecules move through the gel. A gel is a

*Gel electrophoresis
separates substances on the
basis of the size of the
molecules.*

network of molecules, and migrating macromolecules have to find a way through narrow winding passages through the gel. Smaller molecules pass through more easily than larger molecules, and the rate of migration increases as molar mass decreases.

**FIGURE 2.2A**
Apparatus for Gel
Electrophoresis

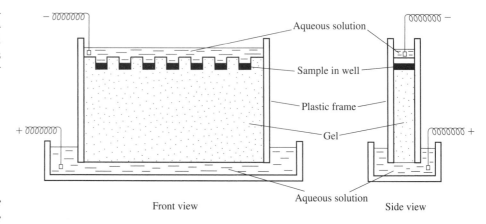

Front view                                                          Side view

*Small molecules move through a gel under the influence of an electric field more rapidly than large molecules.*

After electrophoresis the slab is stained. The components appear as bands [see Figure 2.2B].

**CHECKPOINT 2.2**

**1.** In Figure 2.2B, which band contains the smallest molecules of DNA?

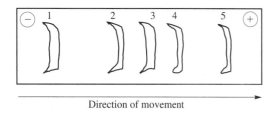

Direction of movement

**FIGURE 2.2B**
Gel Electrophoretogram of Five Samples of DNA

**2.** In the figure which band A or B contains the molecules with the higher positive charge?

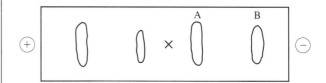

# 2.3  APPLICATIONS OF GEL ELECTROPHORESIS

### 2.3.1  PROTEINS

*Applications of gel electrophoresis include ...*
*... the separation of proteins ...*
*... on the basis of the charge on the molecules ...*
*... or according to the size of the molecules ...*

Gel electrophoresis of proteins can be carried out. Since some proteins have a positive charge and some have a negative charge, the sample must be placed in a central well so that migration can occur in both directions. This separates proteins on the basis of charge.

Proteins can also be separated on the basis of the size of their molecules. They must first be treated with sodium dodecylsulphonate, $C_{12}H_{25}SO_3Na$. This binds to proteins in such a way that it gives each protein molecule the same ratio of charge/mass. As a result the rate of migration increases as molar mass decreases.

## 2.3.2  PURITY

*. . . checking on the purity of substances.*

Gel electrophoresis can be used to check on the purity of substances. If several bands are observed , the substance is impure. If one band is observed, the substance may be pure or it may be a mixture of substances with molecules of the same size. A second method of separation must be tried. If electrophoresis on the basis of charge, as in Figure 2.1A, also fails to produce a separation, the substance must be a pure substance.

## 2.3.3  GENETIC FINGERPRINTING

**DNA profiling**, popularly known as **genetic fingerprinting** is a technique which can identify a person from a sample of their DNA.

*Genetic information is carried in DNA.*

The ability of a living thing to pass genetic information to the next generation is carried in its DNA, deoxyribonucleic acid. DNA is located in the chromosomes which are present in the nuclei of cells . DNA has the famous **double helix** structure. Each strand consists of a chain of molecules of the sugar deoxyribose joined by phosphate groups. Each sugar molecule has a base attached to it, which may be adenine, thymine, cytosine or guanine. The strands are joined together by hydrogen bonds between pairs of bases. Adenine always pairs with thymine, and cytosine always pairs with guanine. [See Figure 2.3A and for more detail see *ALC*, §4.7.3, *Biochemistry and Food Science\**, §5.2 or an A-level Biology text.]

**FIGURE 2.3A**
The Structure of DNA
(S = the sugar
deoxyribose,
P = phosphate,
A = adenine,
T = thymine,
C = cytosine,
G = guanine)

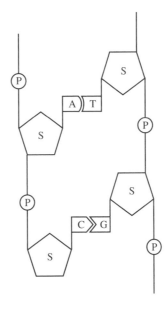

*DNA contains sequences of base pairs called minisatellites, which repeat in a pattern that is unique in each individual.*

In 1984, Alec Jeffreys discovered that human DNA contains sequences of base pairs (10–15 base pairs) which do not carry genetic instructions. The sequences are repeated several times to form a band in the DNA, which Jeffreys called a **minisatellite**. He found that the minisatellites were repeated many times in different parts of the DNA. The pattern of repeats is unique in each individual. Each person inherits half of them from his father and half from his mother [see Figure 2.3B]. The only two people with the same minisatellite sequence are identical twins. For this reason the pattern of

*\*By E.N. Ramsden, published by Stanley Thornes.*

minisatellites is called a **DNA fingerprint**. In unrelated people, one in 4 bands (0.25) match another person's DNA fingerprint. If there are 10 bands in a DNA fingerprint, the chance of all of these matching is $(0.25)^{10}$ or about one in a million.

**FIGURE 2.3B**
Patterns of Minisatellites

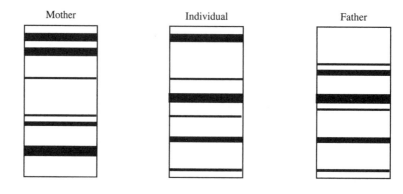

Mother          Individual          Father

*DNA profiling or genetic fingerprinting. . .*
*. . . involves separating DNA from a sample of body tissue . . .*

The first step in obtaining a DNA fingerprint is to obtain a sample of tissue and extract DNA from the nuclei of the cells. Then the DNA is cut at certain points by **restriction enzymes**. Many bacteria make restriction enzymes which they use to protect themselves from foreign DNA, such as invading viruses. Each restriction enzyme recognises a certain sequence of 4–6 base pairs and cuts the DNA at this point. Many restriction enzymes have been obtained from bacteria, and over a hundred are sold commercially.

*. . . cutting it into fragments with restriction enzymes . . .*

The fragments of DNA are separated by **gel electrophoresis** (as in Figures 2.2A and B). The electrophoretogram is invisible at this stage. The gel is covered by a nylon membrane and paper towels, which exert a capillary action and draw the DNA up into the nylon membrane.

*. . . separating the fragments by electrophoresis . . .*
*. . . labelling them with $^{32}P$ . . .*

The sequences must be made visible. The DNA is labelled with radioactive phosphorus-32. This is done by means of **radioactive DNA probes** which bind to complementary bases. An X-ray film is placed on the membrane and the radioactive DNA fragments cause fogging of the film. The pattern of bands is called a **DNA profile** or a **DNA fingerprint**. It can be described by measuring the $R_F$ values of the different bands.

*. . . so that they form an image on an X-ray film . . .*
*. . . an image called a genetic fingerprint.*

A DNA probe is made by genetic engineering techniques [see for example, *Biochemistry and Food Science*, § 5.8 or an A-level Biology text]. A piece of DNA is cut with restriction enzymes, spliced into a ring and introduced into a bacterium. The bacterium reproduces rapidly and reproduces the foreign DNA as it does so. Then the rings of DNA are extracted and a radioactive label is attached by substituting radioactive $^{32}P$ for $^{31}P$ in the phosphate groups.

Genetic fingerprinting can be used to identify whether a man and woman are the parents of a child. In Argentina, during the military regime, many children were taken from their parents and adopted by members of the military junta. Since the return of democratic government, a number of abducted children have been identified and reunited with their families by means of genetic fingerprinting. The technique is also used in cases of disputed paternity to establish who is the father of a child [see Checkpoint 2.3].

*Genetic fingerprinting is used to identify the parents of a child . . .*

Forensic scientists use genetic fingerprints in cases of rape. DNA can be obtained from a vaginal swab taken from the victim and compared with that of a suspect [see Checkpoint 2.3]. In murder cases and assault cases the attacker may leave some of his own blood at the scene of the crime and this can be analysed and compared with the genetic fingerprints of suspects. Matters are not always straightforward. Genetic

*. . . and to identify
criminals.*
fingerprinting was used in the O.J. Simpson trial. Simpson's blood was identified at the scene of the murder, and blood found in his car and at his home was identified as that of the victims. The defence lawyers were able to convince the jury that Simpson's blood had been planted at the scene of the murder by corrupt police officers, who took some of the blood from a sample which Simpson had given for comparison, and that the police had also planted smears of the victims' blood at his house.

### CHECKPOINT 2.3

1. The diagram represents DNA fingerprints. Who is the baby's father?

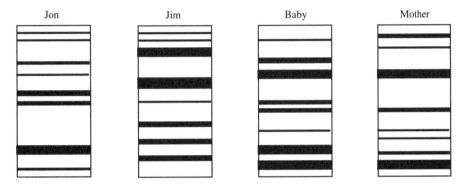

2. The diagram shows the DNA fingerprints of two suspects and a rape victim. Who is the rapist?

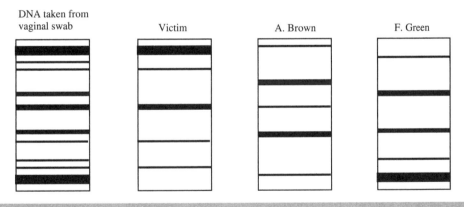

## 2.4 THE STRUCTURE OF PROTEINS

Chromatography and electrophoresis play an important part in the elucidation of the structure of proteins.

### 2.4.1 PURIFICATION

*A protein must be obtained
pure before analysis.
A protein is hydrolysed to a
mixture of amino acids.*

The first step in finding the structure of a protein is to obtain the protein in a pure state. It is precipitated from solution by adding a salt, e.g. ammonium sulphate. A protein has an isoelectric point, a pH at which the ionisation of basic groups and the ionisation of acidic groups are balanced. Most proteins have minimum solubility near the isoelectric point, so precipitation is carried out at this pH. The precipitate is collected by centrifugation.

The next step is to find out which amino acid groups are present. Hydrolysis of the peptide bonds is carried out by heating the protein with $6 \, mol \, dm^{-3}$ hydrochloric acid at $100 \, °C$ to $120 \, °C$ for 10–24 hours. The amino acids which have been formed by hydrolysis are then separated by one of the following methods.

## 2.4.2   SEPARATION

### PAPER CHROMATOGRAPHY

*The mixture can be separated by one of a number of techniques . . .*

*. . . paper chromatography . . .*

The method has been described in § 1.3. The developing solvent is usually butan-1-ol, ethanoic acid and water. The chromatogram may be run in two dimensions [§ 1.3] and is sprayed with ninhydrin detecting reagent to reveal the amino acids. They can be identified from their $R_F$ values.

### THIN LAYER CHROMATOGRAPHY

*. . . TLC . . .*

Often the chromatogram is run in two dimensions. [See § 1.4 and Figure 1.4B.]

### ION EXCHANGE CHROMATOGRAPHY

*. . . ion exchange chromatography, which can be automated . . .*

See § 1.7.4.

### ELECTROPHORESIS

*. . . and electrophoresis.*

The amino acids migrate towards the positive or negative electrode depending on whether they are present as $H_2NCHRCO_2^-$ or as $H_3\overset{+}{N}CHRCO_2H$. The speed at which they travel depends on the ratio of charge/mass [see §§ 2.1, 2.2]. The location of the amino acids is shown by, for example, spraying with ninhydrin.

## 2.4.3   THE SEQUENCE OF AMINO ACIDS

Amino acid analysis is the first step in elucidating the structure of a protein. Proteins have relative molecular masses of 5000 to 36 000. A molecule may contain more than one polypeptide chain, and an individual polypeptide chain contains 100–300 amino acid residues. After finding out which amino acid residues are present, the biochemist sets about finding out the order in which several hundred amino acid residues are linked by peptide bonds to form the polypeptide chain.

*The N-terminus can be identified by chromatography.*

The biochemist finds the identity of the **N-terminus** (the amino acid residue with a free —$NH_2$ group at the end of the chain) by using **Sanger's reagent** [see § 2.4.4], 1-fluoro-2,4-dinitrobenzene, which reacts with an amino group to form a yellow compound.

$$O_2N-\!\!\!\bigcirc\!\!\!-F + H_2N\!\!-\!\!Protein \longrightarrow O_2N-\!\!\!\bigcirc\!\!\!-\overset{\overset{\displaystyle H}{|}}{N}\!\!-\!\!Protein + HF$$
$$\qquad\quad NO_2 \qquad\qquad\qquad\qquad\qquad\qquad\quad NO_2$$

After treatment with Sanger's reagent the protein is hydrolysed, then chromatography comes into play again to separate the amino acids. The terminal amino acid can be identified because it moves as a yellow spot.

The sequence of amino acids in the polypeptide chain is worked out by partially hydrolysing the protein to give small peptide fragments. The sequence of amino acids in the peptide fragments is worked out by further hydrolysis and analysis. By looking for overlaps between sequences of amino acid groups in the peptide fragments, the biochemist gradually assembles a picture of the whole chain. For example, the results of an analysis might be:

*The sequence of amino acids is worked out by conducting partial hydrolyses and studying the peptide fragments.*

N-terminus: His

| Fragments obtained by hydrolysis 1: | Asp Tyr Glu Leu Arg |
| | His Lys |
| | Gly Ala |
| Fragments obtained by hydrolysis 2: | His Lys Asp Tyr |
| | Glu Leu Arg Gly Ala |

From these results can be deduced the sequence:

His Lys Asp Tyr Glu Leu Arg Gly Ala.

You can imagine the meticulous labour involved in completing the sequence for a protein consisting of several hundred amino acids. This sequence is called the **primary structure** of the protein.

### 2.4.4  INSULIN

One of the spectacular applications of separation techniques was the determination of the structure of the hormone **insulin**. Insulin is a protein of relative molecular mass 5700. The task of elucidating its structure was undertaken by F. Sanger of Cambridge University in 1944. The methods of chromatography and electrophoresis were not well developed at that time, and he had to improve techniques as the work progressed. After hydrolysing the protein, he detected and identified 17 amino acids in the hydrolysate. He showed that these were present as 51 amino acid residues in the insulin molecule. He then worked out the sequence in which the 51 amino acid groups occur, developing the use of **Sanger's reagent** en route. His work gained him the Nobel Prize in 1959.

*Chromatography played a big part in the elucidation of the structure of insulin.*

The crystallographer Dorothy Hodgkin took the research further. She had first taken X-ray diffraction photographs of insulin in 1935. After Sanger established the primary structure, she was able to use X-ray diffraction patterns to work out the three-dimensional structure of the molecule. This she achieved in 1970.

### QUESTIONS ON CHAPTER 2

**1.** Describe how electrophoresis is used in the study of proteins. Explain the principles involved.

**2.** (a) Explain the term 'genetic fingerprinting'.

(b) Outline the techniques involved and the role of electrophoresis.

(c) Give two examples of the use of genetic fingerprinting.

**3.** (a) Explain the principle of gel electrophoresis.

(b) Give two examples of mixtures that can be separated by this method.

**4.** Compare the methods that can be used for the separation of amino acids. Comment on their importance in the elucidation of protein structure.

# 3

# LIGHT

## 3.1. WHITE LIGHT

When Isaac Newton (1642–1727) shone a beam of sunlight through a glass prism, he obtained a rainbow of colours from red to violet. The rainbow of colours is composed of visible light of all wavelengths and is called a **continuous spectrum**.

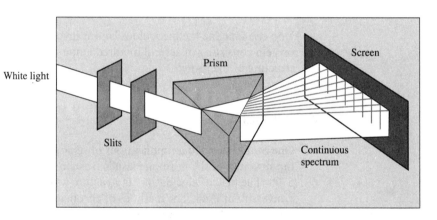

*Visible light consists of light of many different wavelengths. It is called white light.*

Visible light, which consists of light of many different wavelengths, is called **white light**. The wavelengths [Figure 3.1B] range from 430 nm for violet light to 640 nm for red light. Beyond the limits of the visible spectrum are infrared light with wavelength greater than 650 nm and ultraviolet light with wavelength less than 400 nm (1 nanometre, 1 nm = $10^{-9}$ m).

**Wavelength** [see Figure 3.1B] is given the letter $\lambda$ (lambda) and the unit m or nm. **Frequency** (the number of waves in a certain time interval) is given the letter $v$ (nu) and the unit Hertz, Hz, $s^{-1}$. Sometimes the term **wavenumber** measured in $cm^{-1}$ is used to describe a radiation; wavenumber = 1/wavelength in cm. For any wave motion, **velocity** $c$ is given by

*Wavelength $\lambda$, frequency $v$, and velocity of light c, are related: c = $\lambda v$.*

$$c = \lambda v$$

and depends on the medium through which the wave moves. The frequency of a radiation remains constant, but its wavelength changes in different media, thus changing its velocity.

A spectrum can be viewed by an instrument called a spectroscope [Figure 3.1C].

**FIGURE 3.1B**
Wave Motion

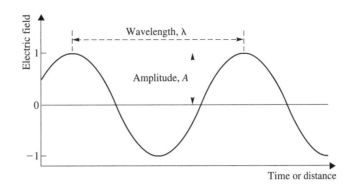

**FIGURE 3.1B**
Wave Motion

**FIGURE 3.1C**
A Spectroscope

1 Light from a source

5 The focusing lens produces a series of coloured images of the slit on a screen.

6 Screen at focal plane.

*White light can be split up into a spectrum of colours ...*
*... and viewed by a spectroscope.*

2 Light is formed into a beam by the slit.

3 A collimating lens forms a parallel beam.

4 The prism disperses the beam into light of different colours.

## 3.2 COLOUR

When white light falls on a block of glass, light of all wavelengths in the visible region of the spectrum is transmitted by (passes through) the glass. The glass object is transparent and appears colourless [see Figure 3.2A(a)]. When visible light falls on an **opaque** object – one which reflects light of all wavelengths in the visible spectrum – the object appears white [see Figure 3.2A(b)].

**FIGURE 3.2A**
(a) A Transparent Object
(b) An Opaque Object

(a)  White light

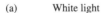

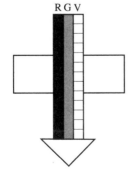

*A transparent object transmits light of all wavelengths in the visible region of the spectrum. An opaque object reflects light.*

(b)  White light

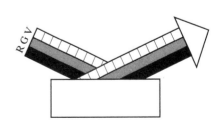

Light of all wavelengths of visible light is transmitted.
The object is transparent.
It appears colourless.

The object is opaque – it does not allow light to pass through.
It reflects visible light of all wavelengths and appears white.

An object may absorb light of a particular wavelength and transmit the rest of the spectrum. The colour of the object which the eye perceives is the result of the combination of the transmitted wavelengths of light. The colour is said to be **complementary** to the colour of light absorbed that is, the colour of the transmitted light and the colour of the absorbed light together add up to white light.

**FIGURE 3.2B**
(a) A Yellow-green Object
(b) A Blue-green Object

*A coloured object absorbs light of some wavelengths and transmits the rest of the spectrum.*

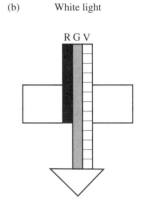

The object absorbs violet light. The wavelengths which it transmits together appear yellow-green.

The object absorbs red light. The wavelengths which it transmits together appear blue-green.

The sets of complementary colours are shown in the **colour wheel** [Figure 3.2C]. Complementary colours, e.g. blue and orange, appear opposite to one another.

**FIGURE 3.2C**
A Colour Wheel

*The wavelengths absorbed and the wavelengths transmitted are complementary . . .
. . . as illustrated by the colour wheel.*

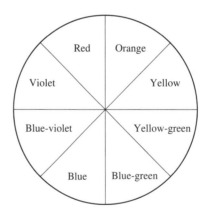

The wavelength and the frequency of light of different colours is shown in Figure 3.2D. The velocity of light, $c$, is related to wavelength, $\lambda$ and frequency, $v$:

$$c = \lambda v$$

**FIGURE 3.2D**
Wavelength, Frequency and Colour of Light

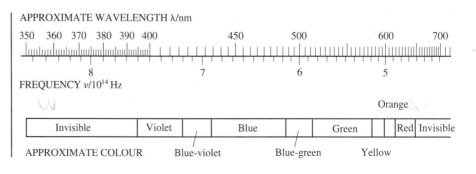

**1.** What colour are objects which absorb light of the following colours?

(*a*) violet, (*b*) red, (*c*) blue, (*d*) yellow, (*e*) yellow-green, (*f*) blue-green.

**2.** What range of wavelengths corresponds to (*a*) red light (*b*) green light?

**3.** What range of wavelengths is being absorbed by an object which appears (*a*) violet (*b*) blue?

## 3.3  ABSORPTION SPECTRA

*A substance may absorb light in the visible part of the spectrum.*

All atoms and molecules absorb light of certain wavelengths. A substance may absorb light of wavelengths in the visible part of the spectrum. Then when white light is passed through the substance black lines appear in the spectrum of colours where light of some wavelengths has been absorbed. The pattern of wavelengths absorbed by a substance is called its **absorption spectrum**. It can be used to identify a substance.

**FIGURE 3.3A**
An Absorption Spectrum

*The pattern of wavelengths absorbed . . .*
*. . . black lines in a spectrum of colours . . .*
*. . . is called the absorption spectrum.*

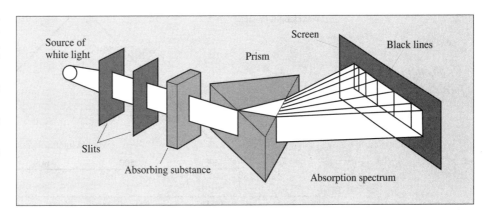

## 3.4  EMISSION SPECTRA

If atoms and molecules are excited sufficiently (given sufficient additional energy), e.g. by heating to a high temperature, they emit light of certain wavelengths. Figure 3.4A shows a discharge tube containing a gaseous element. The observed spectrum consists of a number of coloured lines on a black background. The spectrum is called an **atomic emission spectrum or line spectrum**.

**FIGURE 3.4A**
An Atomic Emission Spectrum or Line Spectrum

*When substances are excited sufficiently they emit light of certain wavelengths . . .*
*. . . as coloured lines on a black background . . .*

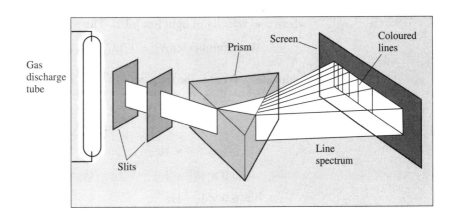

All substances give emission spectra when they are excited in some way, by the passage of an electric discharge or by a flame. The atomic emission spectra of elements are in the visible and ultraviolet region of the spectrum. When sodium or a sodium compound is put into a flame it emits light with a wavelength of 590 nm and colours the flame yellow. A tube of hydrogen gas which has been excited by an electric discharge glows a reddish-pink colour.

*. . . the atomic emission spectrum . . .*
*. . . or line spectrum.*

## 3.5 THE ELECTROMAGNETIC SPECTRUM

Visible light is one of the forms of **electromagnetic radiation** [see Figure 3.5A].

**FIGURE 3.5A**
Electromagnetic
Radiation

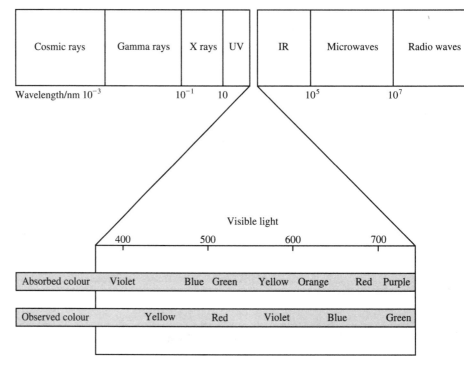

*Visible light is one of the forms of electromagnetic radiation.*

In a vacuum electromagnetic radiation moves at the speed of light, $c$ ($3.00 \times 10^8 \, \text{m s}^{-1}$). Radiation which spans a very narrow band of wavelengths is called **monochromatic radiation**. The wavelength $\lambda$ and frequency $v$ of electromagnetic radiations are related by the expression

*Frequency and wavelength are related.*

$$c = \lambda v$$

where $c$ = velocity of light [see § 3.2]. The wavenumber [see § 3.2] is given by

$$\text{Wavenumber/cm}^{-1} = 1/(\lambda/\text{cm})$$

**Example**    What is the frequency of a microwave beam which has a wavelength of 0.50 cm?

**Solution**

$$v = c/\lambda = 3.0 \times 10^8 \, \text{m s}^{-1}/0.50 \times 10^{-2} \, \text{m}$$

$$= 6.0 \times 10^{10} \, \text{s}^{-1}$$

$$= 6.0 \times 10^{10} \, \text{Hz}$$

*Velocity = wavelength × frequency.*

**Example**   What is the wavelength of an X-ray beam with a frequency of $1.0 \times 10^{18}$ Hz?

**Solution**

$$\lambda = c/v = 3.0 \times 10^8 \text{ m s}^{-1}/1.0 \times 10^{18} \text{ s}^{-1}$$

$$= 3.0 \times 10^{-10} \text{ m}$$

$$= 0.30 \text{ nm}$$

**Example**   What is the wavenumber of an infrared peak at $3.57\,\mu\text{m}$?

**Solution**

*Wavenumber = 1/(Wavelength/cm)*

$$\text{Wavenumber} = 1/\lambda = 1/(3.57 \times 10^{-4} \text{ cm})$$

$$= 2800 \text{ cm}^{-1}$$

**NOTE ON PREFIXES**

Prefixes for units are used to indicate multiples and submultiples of units, e.g. 1 micrometre, $1\,\mu\text{m} = 1 \times 10^{-6}$ metre, $1 \times 10^{-6}$ m. Prefixes you will meet here are:

pico, p $= 10^{-12}$          kilo, k $= 10^3$

nano, n $= 10^{-9}$          mega, M $= 10^6$

micro, $\mu = 10^{-6}$          giga, G $= 10^9$

milli, m $= 10^{-3}$          tera, T $= 10^{12}$

======================== **CHECKPOINT 3.5** ========================

**1.** What colour are solutions A, B and C in Figure 3.5B? Explain how you arrive at your answers.

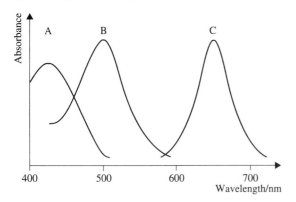

**FIGURE 3.5B**
Absorption Spectra

**2.** Refer to Figure 3.2D. Which light has (*a*) the longest wavelength, (*b*) the highest frequency?

**3.** Refer to Figure 3.5A. Which radiation has (*a*) the longest wavelength, (*b*) the highest frequency?

**4.** (*a*) An infrared spectrum covers the wavelength range of 3.0–15 $\mu$m. Express the range in (i) wavenumbers, (ii) hertz.

(*b*) A visible–ultraviolet spectrum covers the wavelength range of 200–3000 nm. Express the range in (i) wavenumbers, (ii) hertz.

## 3.6  THE HYDROGEN SPECTRUM

*Through a spectrometer, the hydrogen emission spectrum is seen to consist of series of lines.*

Viewed through a spectrometer, the emission spectrum of hydrogen is seen to be a number of separate sets of lines or **series** of lines. These series of lines are named after their discoverers, as shown in Figure 3.6A. The Balmer series, in the visible part of the spectrum, is shown in Figure 3.6B.

| FIGURE 3.6A |
|---|
| The Hydrogen Spectrum |

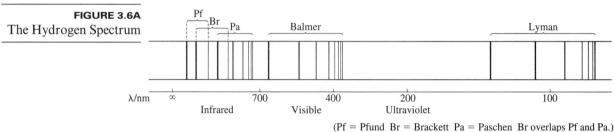

(Pf = Pfund  Br = Brackett  Pa = Paschen  Br overlaps Pf and Pa.)

| FIGURE 3.6B |
|---|
| The Balmer Series of Hydrogen |

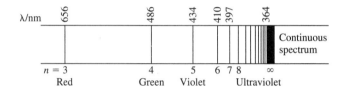

*In each series, the lines become closer together as the frequency increases until at high frequency the lines coalesce.*

In each series, the intervals between the frequencies of the lines become smaller and smaller towards the high frequency end of the spectrum until the lines run together or **converge** to form a **continuum** of light.

### 3.6.1  QUANTISED ENERGY

*Planck theorised that energy is quantised. Bohr suggested that electrons can have only certain amounts of energy ...*

Why do atomic spectra consist of **discrete** (separate) lines? Why do atoms absorb or emit light of certain frequencies? Why do the spectral lines converge to form a continuum? In 1913, Niels Bohr (1885–1962) put forward his picture of the atom to answer these questions. Bohr referred to Max Planck's recently developed **quantum theory**, according to which energy can be absorbed or emitted in certain amounts, like separate packets of energy, called **quanta**. Bohr suggested that an electron moving in an orbit can have only certain amounts of energy, not an infinite number of values: its energy is **quantised**. The energy that an electron needs in order to move in a particular orbit depends on the radius of the orbit. An electron in an orbit distant from the nucleus requires higher energy than an electron in an orbit near the nucleus. If the energy of the electron is quantised, the radius of the orbit also must be quantised.

*... and their orbits can have only certain radii.*

There is a restricted number of orbits with certain radii, not an infinite number of orbits.

An electron moving in one of these orbits does not emit energy. In order to move to an orbit farther away from the nucleus, the electron must absorb energy to do work against the attraction of the nucleus. If an atom absorbs a **photon** (a quantum of light energy), it can promote an electron from an inner orbit to an outer orbit. If sufficient photons are absorbed, a black line appears in the absorption spectrum.

According to the quantum theory, the energy contained in a photon of light of frequency $v$ is $hv$, $h$ being Planck's constant ($6.626 \times 10^{-34}$ J s). For an electron to move from an orbit of energy $E_1$ to one of energy $E_2$, the light absorbed must have a frequency given by **Planck's equation**:

*Electrons which absorb photons move to higher orbits.*

$$hv = E_2 - E_1$$

*Electrons which fall to lower orbits emit photons of light.*

The emission spectrum arises when electrons which have been excited (raised to orbits of high energy) drop back to orbits of lower energy. They emit energy as light with a frequency given by Planck's equation. [See Figure 3.6C.]

**FIGURE 3.6C**
The Origin of Spectral Lines

*Planck's equation gives the frequencies of light emitted.*

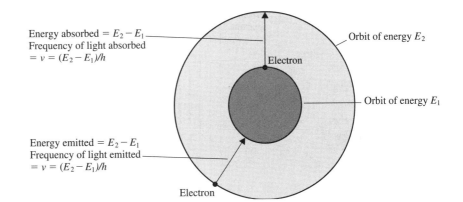

Energy absorbed $= E_2 - E_1$
Frequency of light absorbed
$= v = (E_2 - E_1)/h$

Orbit of energy $E_2$

Electron

Orbit of energy $E_1$

Energy emitted $= E_2 - E_1$
Frequency of light emitted
$= v = (E_2 - E_1)/h$

Electron

Bohr assigned **quantum numbers** to the orbits. He gave the orbit of lowest energy (nearest to the nucleus) the quantum number 1. An electron in this orbit is in its **ground state**. The next energy level has quantum number 2 and so on [see Figure 3.6D]. If the electron receives enough energy to remove it from the attraction of the nucleus completely, the atom is **ionised**.

*Bohr gave orbits of different energy different quantum numbers.*

**FIGURE 3.6D**
The Energy Levels at Various Values of the Quantum Number, $n$

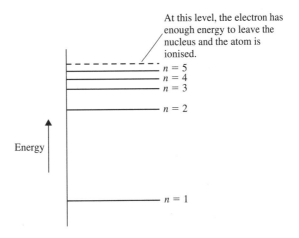

At this level, the electron has enough energy to leave the nucleus and the atom is ionised.

$n = 5$
$n = 4$
$n = 3$

$n = 2$

Energy

$n = 1$

*The hydrogen emission spectrum arises as electrons move from orbits of high quantum number to orbits of lower quantum number.*

Figure 3.6E shows how the lines in the hydrogen emission spectrum arise from transitions between orbits. The Lyman series in the emission spectrum arise when the electron moves to the $n = 1$ orbit (the ground state) from any of the other orbits. The Balmer series arise from transitions to the $n = 2$ orbit from the $n = 3$, $n = 4$ etc. orbits. The Paschen, Brackett and Pfund series arise from transitions to the $n = 3$, $n = 4$ and $n = 5$ orbits from higher orbits.

**FIGURE 3.6E** Energy Transitions in the Hydrogen Atom

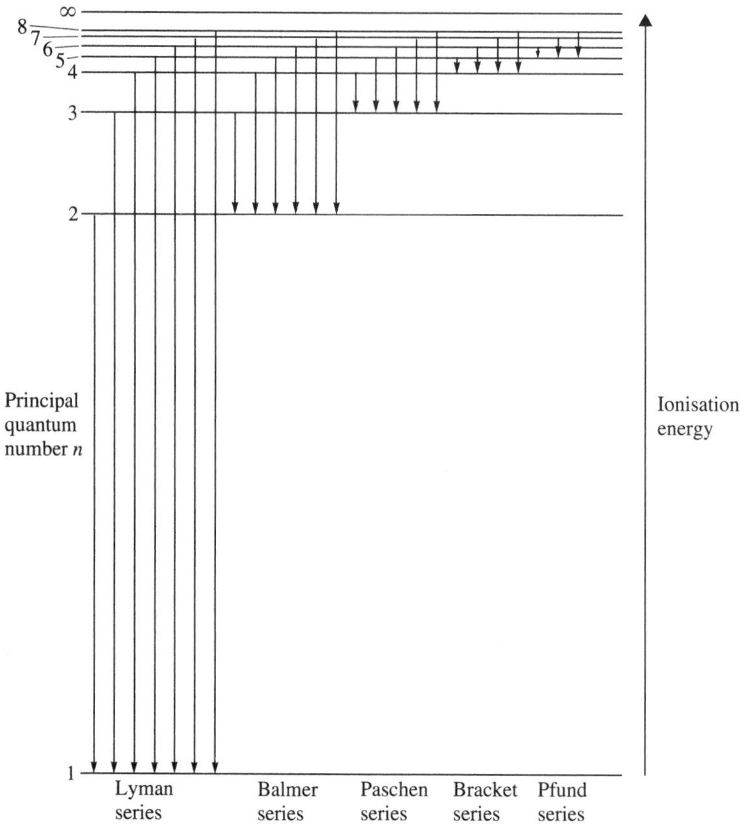

*The frequency of the convergence of spectral lines can be used to give the ionisation energy.*

Principal quantum number $n$

Ionisation energy

Lyman series, Balmer series, Paschen series, Bracket series, Pfund series

In each series of lines, as the frequency increases, each line becomes closer to the previous line until the lines converge, and the spectrum becomes continuous. The Lyman series arises from transitions to the ground state from higher energy levels. The highest frequency lines relate to the highest energy levels. The limit of the Lyman series (the convergence of the lines) corresponds to a transition from the $n = \infty$ orbit (i.e. from an energy level where the electron has escaped from the atom, and the atom has ionised) to the $n = 1$ orbit (the ground state):

*When the lines in the spectrum converge it means that the atom has ionised.*

$$A^+ + e^- \rightarrow A$$

This transition happens when an electron collides with an ion and returns to the ground state. The convergence frequency can be used to find the **ionisation energy** of the atom.

## 3.6.2 DETERMINATION OF IONISATION ENERGY

The first ionisation energy of an element is the energy required to remove one electron from each of a mole of atoms in the gas phase to form a mole of cations in the gas phase:

$$A(g) \rightarrow A^+(g) + e^-$$

*The definition of the first ionisation energy.*

A graphical method can be used to find the value of the ionisation energy from the emission spectrum. The interval between the frequencies of spectral lines becomes smaller and smaller as they approach the continuum.

**1.** The frequencies of the first lines in the Lyman series are measured. If these are $v_1$, $v_2$, $v_3$, $v_4$, etc., the intervals $\Delta v = (v_2 - v_1)$, $(v_3 - v_2)$, $(v_4 - v_3)$, etc., can be calculated.

**2.** A graph of $v$ (the lower frequency) against $\Delta v$ is shown in Figure 3.6F. It can be extrapolated back to $\Delta v = 0$. If there is no interval between lines, this is the beginning of the continuum.

**FIGURE 3.6F**
Finding the Convergence
Frequency by a Graphical
Method

*Finding first ionisation*
*energy by a graphical*
*method ...*

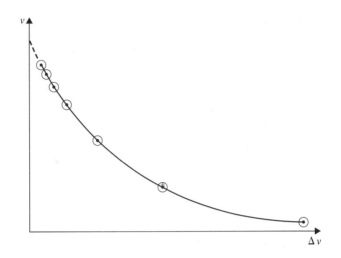

**3.** The value of $v$ at $\Delta v = 0$ is read off and inserted in Planck's equation

$$\Delta E = hv$$

**4.** The value of $\Delta E$ is multiplied by Avogadro's constant to give the first ionisation energy for a mole of atoms.

*... the calculation*

**Example**   The value of the wavelength at the start of the continuum in the sodium emission spectrum is 242 nm. Calculate the first ionisation energy of sodium.

**Method**

Since   $\Delta E = hv$

$\qquad = hc/\lambda$

where $c = 2.998 \times 10^8 \,\mathrm{m\,s^{-1}}$, $h = 6.626 \times 10^{-34} \,\mathrm{J\,s}$ and $L = 6.022 \times 10^{23} \,\mathrm{mol^{-1}}$.

Ionisation energy $= L\Delta E$

$\qquad = Lhc/\lambda$

$\qquad = 6.022 \times 10^{23} \times 6.626 \times 10^{-34} \times 2.998 \times 10^8/(242 \times 10^{-9})$

$\qquad = 494\,300 \,\mathrm{J\,mol^{-1}}$

$\qquad = 494 \,\mathrm{kJ\,mol^{-1}}$

Ionisation energy can also be determined by an electrical method which measures the potential difference at which ionisation takes place.

## QUESTIONS ON CHAPTER 3

**1.** Explain the following terms related to wave motion: wavelength, frequency, amplitude, velocity .

**2.** What is the colour of an object which absorbs (*a*) blue light, (*b*) yellow-green light, (*c*) blue-green light?

**3.** (*a*) Explain briefly how an absorption spectrum arises.

(*b*) Why is the absorption spectrum of an element characteristic of that element?

**4.** (*a*) Explain briefly how an emission spectrum arises.

(*b*) What must be done to make an element give an emission?

**5.** (*a*) Explain how the absorption of photons by an atom leads to an absorption spectrum.

(*b*) Explain why the absorption spectrum and the emission spectrum of a substance are complementary.

**6.** Calculate the frequency of each of the following:

(*a*) an infrared absorption peak at 8260 nm

(*b*) an X-ray beam with a wavelength of 0.25 nm

(*c*) a ruby laser beam with a wavelength of 694 nm.

**7.** Calculate the wavelength of each of the following:

(*a*) an airport radio transmitting at 120 MHz

(*b*) an NMR signal at 105 MHz

(*c*) microwave radiation at $1.5 \times 10^{12}$ Hz

(*d*) an X-ray beam at $3.0 \times 10^{17}$ Hz

**8.**

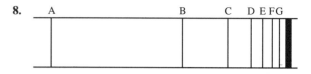

Part of the emission spectrum of atomic hydrogen is shown. Line A is the first of the Lyman series (the series of highest energy).

(*a*) Does the energy increase from left to right or from right to left?

(*b*) Does the frequency increase from left to right or from right to left?

(*c*) Why does the spectrum consist of lines?

(*d*) What do the transitions in a series of spectral lines have in common?

(*e*) Why do the lines become closer together as you read from A to G ?

(*f*) Beyond G the spectral lines run together to become a continuous spectrum. What does this indicate?

**9.** The emission spectrum of atomic hydrogen obtained at high temperatures is more complex and covers a larger part of the spectrum than the emission spectrum obtained at lower temperatures. Suggest a reason for the difference.

# 4

# ATOMIC SPECTROSCOPY

## 4.1 ATOMIC SPECTRA

*Heated elements give out light – the atomic emission spectrum.*

Robert Wilhelm Bunsen, 1811–99, is well known for his burner. He used it in collaboration with another German scientist, Gustav Robert Kirchhoff, 1824–77 to establish the science of **spectroscopy**. Bunsen's burner was used to vaporise samples of the elements, and Kirchhoff's prism spectroscope was used to analyse the light emitted by the hot sample: the **atomic emission spectrum**. They were able to identify several new elements in the Sun by analysing sunlight.

## 4.2 ATOMIC EMISSION SPECTRA

An atomic emission spectrum [see § 3.4] is produced when an atom or ion is excited by the absorption of energy from a hot source and subsequently relaxes to its ground state by emitting a photon of radiation [see Figure 4.2A]. The types of radiation which are sufficiently energetic to cause electronic transitions are ultraviolet, visible and X-ray radiation.

**FIGURE 4.2A**
Excitation and Emission

*The cause is the promotion of an electron to an orbital of higher energy . . .*
*. . . and subsequent relaxation . . .*
*. . . with the emission of light energy.*

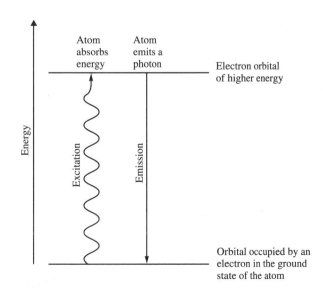

The atomic emission spectrum of hydrogen and its interpretation in terms of the quantum theory are discussed in *ALC*, §§ 2.1 and 2.2.

## 4.3  ATOMIC ABSORPTION SPECTRA

*An element can absorb light by promoting electrons from the ground state to an excited state ...*

An **atomic absorption spectrum** is produced when a gaseous atom or ion absorbs a photon of radiation from an external source [see § 3.3]. Electrons are promoted from the ground state to an excited state – one of several higher energy levels [see § 3.6 and Figure 4.3A]. The atomic absorption spectrum of an element takes the form of a series of narrow peaks (lines). The atomic absorption spectrum and the atomic emission spectrum of an atom or ion are complementary [see § 3.6].

*... giving rise to an atomic absorption spectrum.*

---

**FIGURE 4.3A**
Origin of the Absorption
Spectrum of Sodium

---

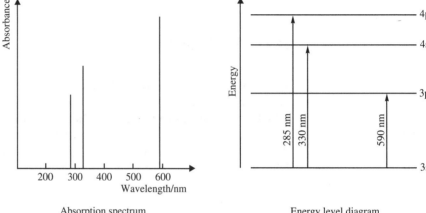

*The atomic absorption spectrum and the atomic emission spectrum are complementary.*

Absorption spectrum                          Energy level diagram

## 4.4  SPECTROSCOPIC METHODS

Atomic spectroscopy gives information about the identity and concentration of <u>atoms</u> in a sample, regardless of how the atoms are combined. This is different from ultraviolet and infrared spectroscopy [Chapters 5 and 6] which give information about the <u>molecules</u> in a sample.

---

**FIGURE 4.4A**
The Components of
Instruments used for
(a) Emission
Spectroscopy,
(b) Absorption
Spectroscopy

---

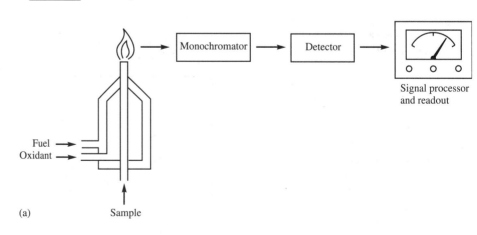

*Atomic spectroscopy gives information about the atoms in a sample, not about molecules (as with visible–ultraviolet and infrared spectra).*

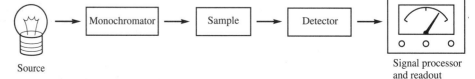

(b)

Spectroscopic studies of atoms and ions can only be carried out in the gaseous state, in which atoms and ions are well separated from one another. In molecular absorption studies [Chapters 5 and 6] the sample is present in a cell that contains a liquid solution or a gaseous sample of the substance. In atomic spectroscopy the sample is present in a flame, a plasma, an arc or a spark. The sample is present as **individual atoms**. Figure 4.4A shows the components of instruments used for atomic spectroscopy.

## 4.5 ATOMISATION

The first step in atomic spectroscopy is **atomisation**, the process by which the sample is vaporised and decomposed in such a way as to produce a gas which consists of atoms and ions. Unless the atomisation step is efficient, the sensitivity and accuracy of the method are poor. Methods of atomisation include flames (1700–3150 °C), plasmas (6000–10 000 °C), electric arcs (4000–5000 °C) and electric sparks (up to 40 000 °C). In **flame atomisation**, an aqueous solution of the sample is dispersed as a fine spray. A stream of fuel and air or oxygen at high velocity carries the spray into the burner [see Figure 4.5A].

**FIGURE 4.5A**
A Burner for Atomic Absorption Spectroscopy

*To give an atomic spectrum, a sample must first be atomised . . .*
*. . . in a flame . . .*
*. . . or in a plasma . . .*
*. . . or in an electric arc or spark . . .*
*. . . and then the atoms are excited.*

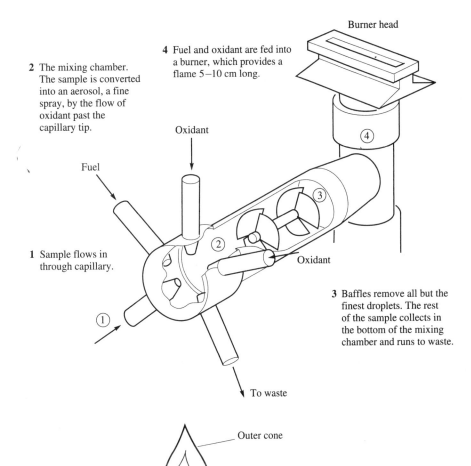

**Burner head**

4 Fuel and oxidant are fed into a burner, which provides a flame 5–10 cm long.

2 The mixing chamber. The sample is converted into an aerosol, a fine spray, by the flow of oxidant past the capillary tip.

Oxidant

Fuel

1 Sample flows in through capillary.

Oxidant

3 Baffles remove all but the finest droplets. The rest of the sample collects in the bottom of the mixing chamber and runs to waste.

To waste

**FIGURE 4.5B**
The Regions in a Flame

*The flame may be produced by the combustion in air of propane, natural gas or hydrogen or the combustion of ethyne in oxygen.*

Outer cone

Inner cone

Base

The solvent evaporates in the base region of the flame [Figure 4.5B]. Fine solid particles are carried into the inner cone, the centre of the flame. Here, gaseous atoms and ions are formed from the solid particles, and **excitation** of atomic spectra takes place. When the atoms and ions reach the outer cone, the outer edge of the flame, they may be oxidised before being dispersed into the atmosphere. The fuel–oxidant mixture flows rapidly through the flame, and only a fraction of the sample is atomised . For alkali metals, low-temperature flames (below 1850 °C) are preferred. They are produced by burning propane or natural gas in air. Hydrogen is burned in air to give flames of 2100 °C and in oxygen to give 2700 °C. Ethyne is burned in oxygen to give flames with temperatures of 2950–3050 °C.

In the flame, all elements ionise to some degree, e.g.

*The spectra of an atom and its ion are different.*

$$Ba \rightleftharpoons Ba^+ + e^-$$

The position of equilibrium depends on the temperature of the flame and on the concentration of barium. The spectra of Ba and $Ba^+$ are different from one another, and in a hot flame the two spectra are excited, one for the atom and one for the ion.

## 4.6   METHODS OF OBTAINING FLAME SPECTRA

### 4.6.1   FLAME EMISSION SPECTRA

Ions of the excited element are the source of radiation, and no external source of radiation is needed. The inner cone of the flame is located in front of the entrance slit of a **monochromator** [see Figure 4.6A and Figure 3.1C]. A monochromator is an instrument that emits light of one wavelength or a narrow band of wavelengths. The output from the exit slit is monitored as the spectrum is scanned by rotating the prism.

**FIGURE 4.6A**
Monochromator

*An atomic emission spectrum needs no external source of radiation: the analyte is the source of radiation.*

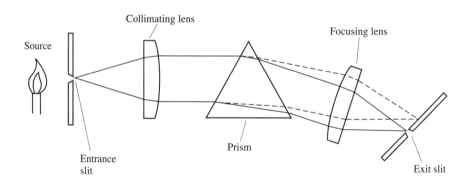

Source

Collimating lens

Focusing lens

Entrance slit

Prism

Exit slit

### 4.6.2   ATOMIC ABSORPTION SPECTRA

Radiation from an external source is passed through the inner cone of the flame, through a monochromator to the surface of a radiation detector. The source of radiation emits lines of radiation that have the same wavelength as the absorption peak of the analyte (the substance to be analysed). This is in contrast with molecular absorption spectroscopy methods which employ a continuous source of radiation.

*An atomic absorption spectrum needs an external source of radiation . . . . . . which emits radiation of the same wavelength as the absorption peak of the analyte.*

The source of radiation is usually a hollow cathode lamp [Figure 4.6B]. As a potential of 300 V is applied across the electrodes, the filler gas, argon (or neon) ionises and $Ar^+$ ions strike the cathode and dislodge metal atoms. Some of the metal atoms are in excited states and emit characteristic wavelengths as they return to the ground state. The wavelength emitted is the wavelength of the absorption peak of the analyte. Hollow cathode lamps for about 40 elements are sold commercially .

**FIGURE 4.6B**
A Hollow Cathode Lamp

*The source of radiation is a hollow cathode lamp.*

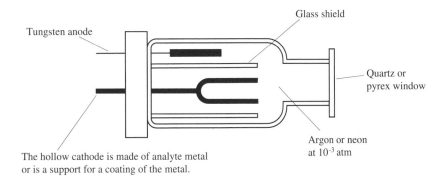

Glass shield

Tungsten anode

Quartz or pyrex window

Argon or neon at $10^{-3}$ atm

The hollow cathode is made of analyte metal or is a support for a coating of the metal.

A **spectrograph** is an instrument which records spectra on a photographic plate or film located along the focal plane of the focusing lens as shown in Figure 4.6C. Spectrographs are used principally for qualitative elemental analysis.

**FIGURE 4.6C**
Double-beam Atomic Absorption Spectrometer

*Spectra may be recorded on a spectrograph.*

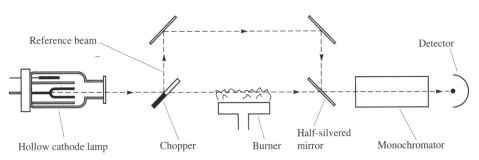

Reference beam

Detector

Hollow cathode lamp

Chopper

Burner

Half-silvered mirror

Monochromator

The chopper is a circle with segments missing. It is motor-driven. Radiation passes through half the time and is reflected the other half.

---

**CHECKPOINT 4.6**

**1.** Explain how the atomic emission spectrum of sodium arises and why it is in the visible region of the spectrum.

**2.** Why does the atomic emission spectrum not give information about the compound of sodium present?

**3.** What item of equipment is required for obtaining an atomic absorption spectrum that is not required for an atomic emission spectrum?

---

## 4.7 SOURCES OF INTERFERENCE IN ATOMIC ABSORPTION SPECTROSCOPY

Two types of interference are sources of error in atomic absorption spectroscopy. They are **spectral interference**, caused either by scattering or by absorption, and **chemical interference**.

### 4.7.1 SPECTRAL INTERFERENCE

#### SPECTRAL INTERFERENCE THROUGH SCATTERING

During atomisation, particles of matter, e.g. metal oxides, may be produced. Such particles scatter incident radiation from the source. This gives a deceptively high value for the absorption of radiation.

## SPECTRAL INTERFERENCE THROUGH ABSORPTION

*A source of error in atomic absorption spectroscopy is spectral interference . . .*
*. . . due to scattering by combustion products . . .*
*. . . and due to absorptions by molecular species which overlap that of the analyte . . .*
*. . . and due to absorption by the combustion products of the analyte . . .*
*. . . and the combustion products of the fuel.*

**1.** A species will interfere if its absorption or emission spectrum overlaps the analyte wavelength. Often the problem can be solved by choosing a different line at which to measure absorption or emission. For example when aluminium is analysed by using the line at 308 nm, a vanadium line at 308 nm interferes, but if the aluminium line at 309 nm is selected there is no interference.

**2.** Combustion may produce molecular products, and these have broad-band absorption which may overlap that of the analyte. The atomic absorption spectrum of barium is in the centre of an absorption band for CaOH; therefore calcium interferes in the analysis of barium. The problem is solved by working at a higher temperature to decompose CaOH.

**3.** Absorbing products may be formed from the fuel–oxidant mixture. A correction can be made by running a blank without adding the sample and subtracting this from the absorption measured with the sample.

### 4.7.2 CHEMICAL INTERFERENCE

*Another source of error is chemical interference by combination of the metal analyte with anions present in the sample . . .*
*. . . a problem solved by . . .*
*. . . raising the temperature . . .*
*. . . or adding a releasing agent . . .*
*. . . or a protective agent.*

**1.** The sample may contain anions that form compounds with the analyte. If calcium is the analyte, its volatility is decreased by the presence of sulphate ions and phosphate ions with which it combines to form non-volatile compounds. The solution to this problem may be to work at higher temperatures to overcome low volatility. Alternatively a **releasing agent** can be added, a substance which has a strong affinity for the anions and prevents them combining with the analyte. For example, strontium could be added to a sample of calcium because phosphate ions and sulphate ions combine with strontium in preference to calcium. Another solution is to add a **protective agent**. This is a substance which forms a stable but volatile compound with the analyte. Edta and other complexing agents are used, e.g. edta forms a stable, volatile complex with calcium and prevents other anions from interfering.

**2.** In flames which use air as oxidant, atoms do not usually ionise. When oxygen is the oxidant, the temperature is higher, and atoms may ionise:

$$M \longrightarrow M^+ + e^-$$

*Ionisation of the analyte leads to a low measurement for absorption . . .*
*. . . and is reduced by adding an ionisation suppressor.*

The spectrum of $M^+$ is different from that of M, and the absorption peak due to M is therefore lowered. This problem can frequently be cured by adding an **ionisation suppressor**, which adds electrons to the flame and encourages the reverse reaction:

$$M^+ + e^- \longrightarrow M$$

Salts of a metal with a low ionisation energy, e.g. potassium, are often used as ionisation suppressors.

## 4.8 CALIBRATION

*The height of the absorption peak is given by the Beer–Lambert Law.*

Atomic absorption methods are used for **quantitative analysis**. The height of an absorption peak at a certain wavelength depends on the amount of absorbing substance present and can be used to measure its concentration. The height of an absorption peak due to an absorbing substance in a non-absorbing solvent is given by the **Beer–Lambert Law**:

$$A = \lg(I_0/I) = \varepsilon c l$$

where $A$ = absorbance, $I_0$ = intensity of incident beam, $I$ = intensity of transmitted beam, $\varepsilon$ = the molar absorption coefficient, which is constant for a particular species (molecule or ion) at a certain wavelength, $c$ = concentration, $l$ = path length, that is the thickness of the cell. The value of $\varepsilon$ must be found for the species at the wavelength being used. Then the measured value of $A$ for a solution of the species will give the concentration of the species.

*Departures from the Beer–Lambert Law occur. In atomic absorption spectroscopy a standard must be run every time an analysis is performed. The standard-addition method compensates for interference.*

Departures from the linear relationship expressed in the Beer–Lambert Law occur, however, and it is impossible to obtain accurate analyses by relying on the assumption that the Beer–Lambert Law is followed. The absorbance of a standard solution must be measured each time an analysis is performed. The **standard-addition method** can be used. Two equal measured volumes of the sample are taken. To one is added the standard – a known amount of the analyte. Then both portions are diluted to the same volume. The absorbance of each solution is measured, and the absorbance due to the sample is calculated.

## 4.9  APPLICATIONS OF ATOMIC SPECTROMETRY

### 4.9.1  ELEMENTAL ANALYSIS

Atomic emission spectrometry is widely used for the determination of half the elements in the Periodic Table. It is especially useful for the determination of sodium, potassium, lithium and calcium in biological fluids and tissues because these elements are difficult to determine by other methods. Instruments are manufactured specifically for the analysis of sodium, potassium and lithium in blood and other biological fluids. The radiation from the flame is split into three beams. Each beam passes into a separate spectrophotometric system, in which a filter transmits the emission line of one of the three elements and absorbs the other two. An internal standard must be used.

**FIGURE 4.9A**
Potassium Emission

*Applications of atomic emission spectroscopy include the analysis of Na, K, Li and Ca in biological fluids ...*

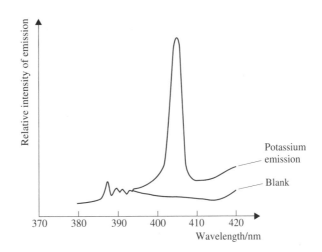

### 4.9.2  FORENSIC SCIENCE

*... forensic science ...*

Forensic scientists may want to compare a paint fragment from a suspect's clothing with paint from the scene of a crime. Since different manufacturers add different elements and dyes to paints, paint samples can be identified by atomic emission spectroscopy. Other methods can also be used. The resins in paints can be identified by infrared spectrometry [Chapter 6], and also by the pyrolysis of paint fragments followed by gas chromatography and mass spectrometry [Chapter 8].

### 4.9.3. FIREWORKS

*...fireworks...*

Fireworks contain elements and compounds described as **pyrotechnics**. They burn in a controlled manner to form solid products (unlike propellants and explosives which form gases) and give out heat and light. The heat given out is sufficient to excite electrons and make the atoms present show their emission spectra. Different fireworks have contents which include calcium compounds for brick-red light, strontium compounds for crimson light, sodium compounds for golden yellow light and barium compounds for green light.

### 4.9.4 POLLUTION MONITORING

*...and pollution monitoring.*

Atomic absorption spectrometry is used to analyse environmental samples for the presence of metals. The sample is aspirated into the flame [see Figure 4.5A]. For some metals, e.g. mercury, an electrothermal method may be used [§4.10.3]. Mercury can be detected down to $10^{-9}$ g by its absorption at 254 nm. **Water pollutants** which are analysed by atomic absorption spectrometry include barium, cadmium, chromium, zinc and lead.

**Water quality** is analysed by measuring the content of calcium and magnesium (for hardness), copper (for plant growth), aluminium (used in water treatment to coagulate colloidal matter), iron, etc.

## 4.10  ADVANCED TECHNIQUES

### 4.10.1  COOLING

*Advanced techniques include spectroscopy at low temperature ...*

Cooling molecules down to low temperatures brings them down to their lowest energy levels. This simplifies spectra which are complex at higher temperatures.

### 4.10.2  LASERS

Lasers can deliver radiation of very closely defined wavelength and can deliver pulses of light lasting a nanosecond ($10^{-9}$ s) enabling spectroscopists to study unstable and very reactive species.

### 4.10.3  ELECTROTHERMAL ATOMISERS

*...and electrothermal absorption spectroscopy.*

Flame atomisation is not very efficient. A large portion of the sample either flows out to waste [Figure 4.5A] or is not completely atomised. The residence time of atoms in the optical flame is brief ($\sim 10^{-4}$ s). Electrothermal atomisers are much more sensitive. A small volume (a few $\mu$l) of sample is evaporated and ashed in an electrically heated cup, then the current is rapidly increased to several hundred amperes, causing the temperature to rise rapidly to 2000–3000 °C. Atomisation occurs in a period of a few milliseconds to seconds. The absorbance of the atomised particles in the region above the electrically heated cup is determined by fitting the electrothermal atomiser in front of the entrance slit of a monochromator (as in Figure 4.6A).

# QUESTIONS ON CHAPTER 4

1. What is meant by the statement that atomic absorption spectroscopy and atomic emission spectroscopy are 'complementary'? How does this state of affairs arise?

2. Briefly describe how a sample is atomised to give rise to an atomic absorption spectrum. Say what happens in each of the three regions of the flame. Name three fuels that are burned to give a flame.

3. An atomic emission spectrum is obtained by locating a flame in front of a monochromator or spectrophotometer. To obtain an atomic absorption spectrum, an additional piece of equipment is needed.

(a) What is the additional piece of equipment?

(b) Briefly describe it.

4. Sources of error in atomic absorption spectrometry are (a) scattering, (b) absorption by other species and (c) chemical interference.
Explain briefly what is meant by each of these, and suggest how they can be overcome.

5. (a) What is the *absorbance* of a species? How is it related to the concentration of the species?

(b) A lead-containing sample of volume $50 \, \mu dm^3$ was analysed by atomic absorption spectrometry and gave an absorbance of 0.075. Lead was added to the sample to a concentration of $6.0 \, \mu g \, dm^{-3}$. The sample to which lead had been added had an absorbance of 0.115. Calculate the concentration of lead in the untreated sample.

6. The sodium content of a cement sample was determined by flame emission spectrometry. For calibration, standards were used; see table. Sample A of cement, 1.000 g, was dissolved in hydrochloric acid, neutralised, made up to $100.0 \, cm^3$, and the emission was measured. Samples B and C were duplicates of sample A; see table.

(a) Plot the value of emission against Na content of the standards.

(b) Calculate the sodium content (% by mass) of the cement.

| Standard/ $\mu g \, Na \, cm^{-3}$ | Emission |
|---|---|
| 0 | 3.0 |
| 20 | 22.0 |
| 40 | 40.0 |
| 60 | 58.0 |
| 80 | 76.0 |

| Sample | Emission |
|---|---|
| A | 35.3 |
| B | 36.4 |
| C | 35.6 |

7. A sample $5.00 \, cm^3$ of blood was treated to precipitate proteins and centrifuged. The solution was brought to pH 3 and extracted with a lead-chelating agent. The extract was drawn into an air–ethyne flame and gave an absorbance of 0.502 at 283.3 nm. Two $5.00 \, cm^3$ portions of standard solutions containing (a) 0.440 ppm lead and (b) 0.600 ppm lead which were treated in the same way gave absorbances of (a) 0.396 (b) 0.599 respectively. Calculate the lead content of the blood sample in ppm (assuming the Beer–Lambert Law is followed).

8. Give three uses of atomic spectroscopy.

# 5

# VISIBLE–ULTRAVIOLET SPECTROMETRY

## 5.1 VISIBLE–UV, IR AND MICROWAVE SPECTRA

The visible–ultraviolet spectrum of benzene is shown in Figure 5.1A.

**FIGURE 5.1A**
Absorption Spectrum of
Benzene

*A visible–ultraviolet spectrum measures the absorption of light in the visible–ultraviolet region of the spectrum in the form of a plot of the intensity of transmitted light against wavelength.*

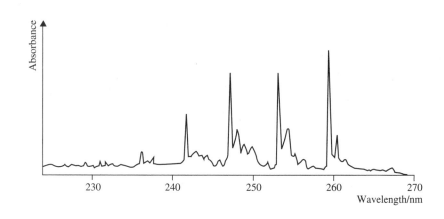

This absorption spectrum is obtained by placing a solution of benzene between a source of visible and ultraviolet (UV) light and the detector. The detector analyses the intensity of the transmitted light relative to the intensity of the incident light for each frequency. Modern instruments scan the visible–UV spectrum automatically and produce a plot of intensity of absorbed energy against wavelength.

*When a molecule absorbs a photon of energy, it may increase . . .*

*. . the electronic energy of the molecule (visible–ultraviolet radiation) . . .*

*. . . or vibrational energy of the molecule (infrared radiation) . . .*

Molecules absorb photons of energy from radiation. The energy of the photon raises the molecule that has absorbed it to an **excited state**. Different types of radiation produce different kinds of excited states. Visible and UV light can excite a molecule to a higher electronic energy state by promoting an electron to an orbital of higher energy [see § 3.6]. In some cases, the bonds between the atoms in the molecule are weakened, as in photochemical reactions [*ALC*, § 25.8.2, 26.3.8]. Infrared radiation is of lower energy than UV radiation. It cannot promote electrons but it can increase the **vibrational energy** of a molecule. The ways in which the molecule $H_2O$ can vibrate are shown in Figure 5.1B.

*. . . or the rotational energy of the molecule (microwave radiation).*

In addition to vibrating and bending, a molecule may rotate. The absorption of **microwave radiation**, which is radiation of low photon energy, raises molecules to higher levels of **rotational energy**.

**FIGURE 5.1B**
Modes of Vibration of the
$H_2O$ Molecule

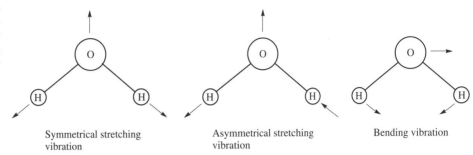

Symmetrical stretching
vibration

Asymmetrical stretching
vibration

Bending vibration

*The absorption of radiation is governed by the relationship . . .*
*. . . ΔE = hv . . .*
*. . . where ΔE = energy absorbed, h = Planck's constant, v = frequency.*

The frequency of the radiation that can be absorbed by an atom or a molecule depends on the difference in energy $\Delta E$ between the ground state and the excited state. It is given by

$$\Delta E = hv$$

where $h$ = Planck's constant , $v$ = frequency of radiation [see § 3.6].

## 5.2  THE ENERGY LEVELS OF A MOLECULE

The biggest differences in energy are between different **electronic energy levels**. Each electronic energy level has its own subset of **vibrational levels** [see Figure 5.2A], and each of these has its own subset of **rotational levels**.

**FIGURE 5.2A**
Energy Levels in a
Molecule

*The electronic, vibrational and rotational energy levels of a molecule are illustrated.*

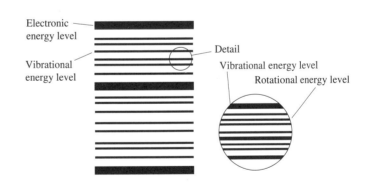

Electronic
energy level

Detail

Vibrational
energy level

Vibrational energy level

Rotational energy level

## 5.3  VISIBLE–ULTRAVIOLET SPECTRA

*Visible–ultraviolet spectra arise from transitions between electronic energy levels which differ in their vibrational and rotational energies to higher electronic energy levels of slightly different energies. Absorption bands, not lines, therefore appear.*

A transition between two electronic energy levels should give rise to a line in an absorption spectrum. In practice it may not be possible to distinguish individual lines. Large molecules have a large number of vibrational and rotational energy levels. These energy levels may be close together. Visible–ultraviolet radiation may raise molecules at many slightly different electronic energy levels to many slightly different higher levels. As a result, sharp lines are replaced by narrow bands – **absorption bands** (as in benzene, Figure 5.1A).

Spectroscopy in the visible and UV regions is carried out in solution. Collisions with solvent molecules can alter individual lines. Figure 5.3A shows how the spectrum of a substance can alter with the solvent.

**FIGURE 5.3A**

UV Spectrum of Ethyl-3-
oxobutanoate in
A, Hexane;
B, Ethoxyethane;
C, Ethanol; D, water

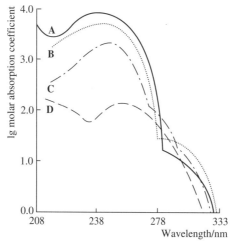

Molar absorption coefficient $= \lg(I_0/I)cl$
$I_0$ and $I$ are the intensities of
the incident and transmitted
light respectively.
$c$ = concentration of solution/mol dm$^{-3}$
$l$ = length of cell/cm

## 5.4 COLORIMETRY

*The absorption of a
solution can be measured
and used to find the
concentration of solute.*

The height of an absorption peak at a certain wavelength depends on the amount of
absorbing substance present and can be used to measure its concentration. A solvent
which does not absorb at the same wavelength should be chosen if possible. The
method compensates for any slight absorption due to the solvent. The height of an
absorption peak due to an absorbing substance in a non-absorbing solvent is given by
the Beer–Lambert Law:

$$A = \lg (I_0/I) = \varepsilon c\, l$$

[See § 4.8 for the symbols.]

**FIGURE 5.4A**

A Colorimeter

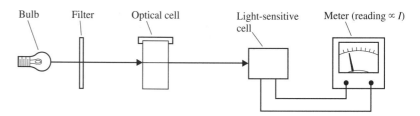

Bulb    Filter    Optical cell    Light-sensitive cell    Meter (reading $\propto I$)

───── **CHECKPOINT 5.4** ─────

**1.** Samples of C and D transmit 25.0% and 75.0%
respectively of the light incident upon them. What are their
absorbances?

**2.** Solutions E and F have absorbances of 0.125 and 0.725
respectively.

(*a*) What percentage of the incident light does each solution
transmit?

(*b*) What is the percentage transmittance of solutions which
have twice the absorbance of E and F?

**3.** A solution contains 4.48 ppm $KMnO_4$. In a 1.00 cm cell
the absorbance of the solution at 520 nm is 0.691. Calculate
the molar absorption coefficient of $KMnO_4$.

## 5.5 FINDING THE FORMULA OF A COMPLEX ION

When solutions of nickel(II) sulphate and edta are mixed, coloured complex ions are
formed. If the formula of the ion is $[Ni(edta)_n]^{x-}$ then the maximum intensity of
colour will be obtained when the two solutions are mixed in the molar ratio of

Ni : edta $= 1 : n$

| TABLE 5.5A | Solution | Volume of NiSO₄/cm³ (0.05 mol dm⁻³) | Volume of edta/cm³ (0.05 mol dm⁻³) |
|---|---|---|---|
| | 1 | 0 | 10 |
| | 2 | 1 | 9 |
| | 3 | 2 | 8 |
| | 4 | 3 | 7 |
| | 5 | 4 | 6 |
| | 6 | 5 | 5 |
| | 7 | 6 | 4 |
| | 8 | 7 | 3 |
| | 9 | 8 | 2 |
| | 10 | 9 | 1 |
| | 11 | 10 | 0 |

*The formula of a complex ion can be found by colorimetry.*

Solutions are made up as shown in Table 5.5A. They are put into a colorimeter [see Figure 5.4A]. The absorbance of each solution is measured. The absorbance is proportional to the concentration of the complex. Figure 5.5A shows a plot of absorbance against the number of the solution. The maximum colour intensity corresponds to the formula $[Ni(edta)]^{2-}$.

**FIGURE 5.5A**
Results of Colorimetry on Ni–edta Complex

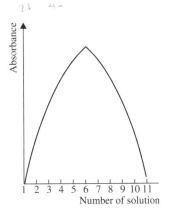

1 2 3 4 5 6 7 8 9 10 11
Number of solution

Absorbance

═══════════════════════ **CHECKPOINT 5.5** ═══════════════════════

1. The visible–UV spectra of solutions of compounds **A**, **B** and **C** are shown below. What colours are **A**, **B** and **C**?

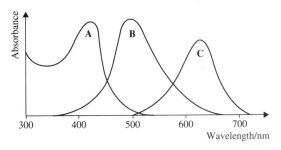

300    400    500    600    700
Wavelength/nm

Absorbance

**FIGURE 5.5B**
The Visible–UV Spectra of **A**, **B** and **C**

2. Volumes of 0.100 mol dm⁻³ $M^{2+}$(aq) and 0.100 ml dm⁻³ **L**(aq) are added as shown in the table. **M** is a transition metal, and **L** is a ligand which forms a complex with $M^{2+}$.

The absorbance of each solution is measured. Use the results shown to obtain the formula of the complex ion.

| Volume of $M^{2+}$(aq)/cm³ | Volume of L(aq)/cm³ | Absorbance |
|---|---|---|
| 0 | 10.0 | 0 |
| 1.0 | 9.0 | 0.293 |
| 2.0 | 8.0 | 0.569 |
| 3.0 | 7.0 | 0.759 |
| 4.0 | 6.0 | 0.793 |
| 5.0 | 5.0 | 0.759 |
| 6.0 | 4.0 | 0.655 |
| 7.0 | 3.0 | 0.465 |
| 8.0 | 2.0 | 0.310 |
| 9.0 | 1.0 | 0.155 |
| 10.0 | 0 | 0 |

# 5.6 TYPES OF SPECTROSCOPIC INSTRUMENTS

*Different types of spectroscopic instruments are described, including ...*
*... the monochromator ...*
*... the spectroscope ...*

A **monochromator** is an instrument that emits light of one wavelength or a very narrow band of wavelengths.

A **spectroscope** [see Figure 3.1C] is an instrument for identifying the elements in a sample that has been excited by a flame or otherwise. It has a monochromator and an eyepiece that allows visual inspection of the emission lines. The eyepiece is movable and the wavelength of a line is determined from the angle between the incident beam and the path of the line to the eyepiece.

A **colorimeter** is an instrument that measures absorption. The human eye is the detector, and a standard is used for comparison.

A **photometer** is an instrument that can be used for absorption, emission or fluorescence measurement with ultraviolet, visible or infrared radiation. A photometer has filters for wavelength selection and a photoelectric device for measuring radiation. Instruments used for absorption measurement with visual radiation are often called colorimeters (although this is not perfectly correct; see above).

*... the colorimeter ...*
*... the photometer ...*
*... the spectrograph ...*
*... the spectrometer ...*

A **spectrograph** records spectra on a photographic plate or film located along the focal plane of a monochromator. In Figure 3.1C the screen at the focal plane would be replaced by a plate or film holder. Spectrographs are used principally for qualitative analysis of elements.

A **spectrometer** has a monochromator which has a fixed slit at the focal plane and a detector. A spectrometer with a phototransducer (which converts light energy into electrical energy) at the exit slit is called a **spectrophotometer**. Spectrometers can be used for absorption, emission and fluorescence measurements.

*... the visible–ultraviolet spectrophotometer ...*

A **visible–ultraviolet spectrophotometer** is a spectrophotometer which measures the absorption of radiation over the visible–UV part of the electromagnetic spectrum. Spectroscopic instruments can be single-beam, double-beam or multi-channel. A **double-beam visible–ultraviolet spectrophotometer** is shown in Figure 5.6A. One beam passes through a cell containing a solution of the sample to a photomultiplier. The second beam passes simultaneously through a cell containing the solvent to a second,

---

**FIGURE 5.6A**
A Visible–Ultraviolet
Spectrophotometer

*... and the double-beam spectrophotometer in which one beam passes through the sample and one through a reference cell containing solvent alone.*

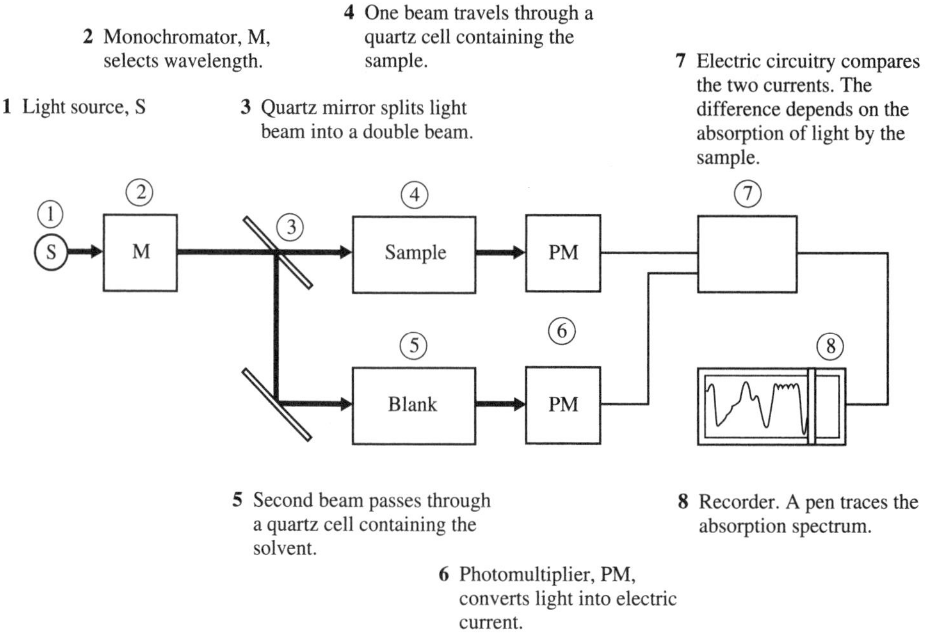

*The cells are made of silica which does not absorb visible–ultraviolet light. The transmitted light gives rise to an electric current.*

**2** Monochromator, M, selects wavelength.

**4** One beam travels through a quartz cell containing the sample.

**7** Electric circuitry compares the two currents. The difference depends on the absorption of light by the sample.

**1** Light source, S

**3** Quartz mirror splits light beam into a double beam.

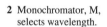

**5** Second beam passes through a quartz cell containing the solvent.

**6** Photomultiplier, PM, converts light into electric current.

**8** Recorder. A pen traces the absorption spectrum.

*The instrument plots absorbance against wavelength.*

matched photomultiplier. The intensities of the two beams are compared and the ratio is displayed by a readout device. In modern instruments the spectrum is plotted automatically as $\log(I_0/I)$ against wavelength. Radiation over the entire visible–ultraviolet range can be scanned in 30 s.

To provide radiation over the whole of the visible–ultraviolet range the instrument needs a hydrogen discharge lamp to cover the UV range and a tungsten lamp to cover the visible range. The cells, which are usually 1 cm$^2$ in cross section and 5 cm high, are made of silica because soda glass absorbs below 365 nm and pyrex glass absorbs below 320 nm.

The instrument is calibrated for wavelength by passing the sample beam through specimens of glass that have known absorption peaks. Calibration of absorption is done by passing the sample beam through filters of known absorbance or through standard solutions.

*For the single-beam spectrophotometer see § 5.11.4.*

**Single-beam spectrophotometers** are described in § 5.11.4.

## 5.7  COLOURED ORGANIC COMPOUNDS

*Coloured substances absorb light in the visible region of the spectrum.*

A substance appears coloured if it absorbs light in the visible region of the spectrum [see § 3.2]. The light energy which it absorbs is used to promote electrons from the ground state to orbitals of higher energy. Most organic compounds absorb in the visible–UV region of the spectrum. The feature which they have in common is the possession of double bonds or triple bonds or non-bonded pairs of electrons (lone pairs). The UV spectra of alkenes, which possess the group C=C, have a peak at 180 nm. It arises from the excitation of an electron in the C=C double bond to a higher energy level. Aldehydes and ketones with the C=O group show a peak at 280 nm which arises from the excitation of an electron in the C=O bond to a higher level. The positions of peaks can be altered within limits by neighbouring groups, e.g. —CH$_3$ and —Cl. Groups such as C=C and C=O which give a substance a distinctive absorption peak are called **chromophores** (Greek for 'colour-bringer'). A substance which contains a **chromophore** is called a **chromogen**. The mechanism by which chromophores absorb light must depend on the molecular orbitals involved, as discussed below.

*Organic compounds absorb in the visible–ultraviolet region if they contain groups such as >C=C< or >C=O or lone pairs of electrons.*

*Such groups are called chromophores . . .*

Some chromophores are:

$$\begin{array}{ccccc} \diagdown\!\!\diagup\ \ \ & \diagdown\ \ \ & & \diagdown\ \\ \diagup\mathrm{C}\!=\!\mathrm{C}\diagdown & \diagup\mathrm{C}\!=\!\mathrm{O} & -\mathrm{CONH_2} & \diagup\mathrm{C}\!=\!\mathrm{N}\!- \\ \\ -\mathrm{C}\!\equiv\!\mathrm{C}\!- & -\mathrm{COCl} & -\mathrm{NO_2} \end{array}$$

*. . . and the compounds that contain them are called chromogens.*

## 5.8  BONDING AND ANTIBONDING ORBITALS

*Atomic orbitals combine to form molecular orbitals . . .*
*. . . which may be bonding orbitals . . .*
*. . . or antibonding orbitals at a higher energy level.*

The **wave theory of the atom** [*ALC*, § 2.3] treats electrons as having the properties of particles and also the properties of waves. Electrons are considered in terms of electron densities that extend over the whole atom. When atoms combine in a covalent bond, the **atomic orbitals** combine to form **molecular orbitals** [*ALC*, § 5.2]. The wave theory allows two possible combinations. Figure 5.8A shows how two 1s atomic orbitals combine to form a **bonding orbital**. A different combination of atomic orbitals produces a different orbital which is called an **antibonding orbital**. The bonding orbital is at a lower energy level than the separate atomic orbitals, and the antibonding orbital is at a higher energy level than the separate atomic orbitals.

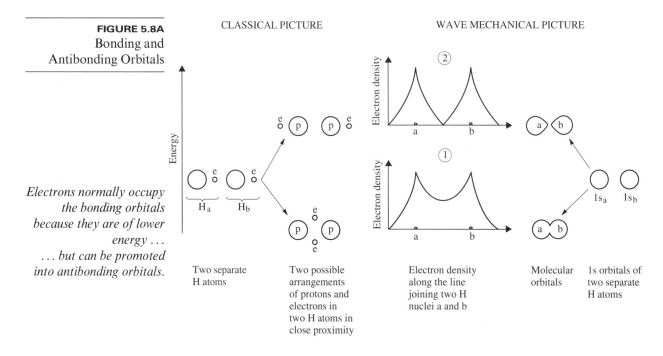

CLASSICAL PICTURE                    WAVE MECHANICAL PICTURE

**FIGURE 5.8A**
Bonding and
Antibonding Orbitals

*Electrons normally occupy
the bonding orbitals
because they are of lower
energy ...
... but can be promoted
into antibonding orbitals.*

Two separate
H atoms

Two possible
arrangements
of protons and
electrons in
two H atoms in
close proximity

Electron density
along the line
joining two H
nuclei a and b

Molecular
orbitals

1s orbitals of
two separate
H atoms

**1** Note the high electron density between the nuclei, which reduces the repulsion between the nuclei and permits bonding. This is a **bonding molecular orbital**.

**2** The electron density is low in the internuclear region. It does little to reduce repulsion between the nuclei. This is an **antibonding molecular orbital**.

Electrons normally occupy the lowest energy molecular orbitals available to them in a molecule. The antibonding orbital is therefore unoccupied in the ground state. An electron must receive energy to promote it from a bonding orbital into an antibonding orbital. The difference in energy between bonding orbitals and antibonding orbitals depends on the type of orbital. There are three types of molecular orbitals. These are:

*Molecular orbitals can be $\sigma$
orbitals, $\pi$ orbitals and
n orbitals.*

- $\sigma$ orbitals formed from atomic orbitals of the s, sp, $sp^2$ and $sp^3$ type [see *ALC*, §5.2.7]

- $\pi$ orbitals, e.g. in ethene and benzene [see *AlC*, §5.2.7]

- n-orbitals occupied by lone pair electrons in non-bonding orbitals on atoms such as nitrogen and oxygen within the organic molecule [see *ALC*, §5.1.2, 5.1.3, 5.1.5].

The carbonyl group possesses all three types of orbital [Figure 5.8B].

**FIGURE 5.8B**
Orbitals in the Carbonyl
Group

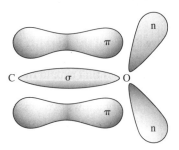

**FIGURE 5.8C**
Permitted Electronic
Transitions

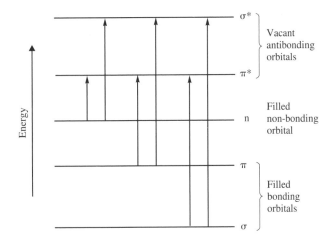

**FIGURE 5.8C**
Permitted Electronic
Transitions

*Promotion of an electron
can occur from $\pi \longrightarrow \pi^*$,
from $n \longrightarrow \pi^*$ and
$n \longrightarrow \sigma^*$ with the
absorption of energy and the
production of an absorption
spectrum in the visible–
ultraviolet region.*

It is possible for electrons in the carbonyl group to be excited to higher orbitals. When an electron is excited, an electron from one of the filled orbitals, $\sigma$, $\pi$ or n, is promoted to an antibonding orbital $\sigma^*$ or $\pi^*$.

Figure 5.8C shows the transitions which are possible. The transitions $\pi \longrightarrow \pi^*$, $n \longrightarrow \pi^*$ and $n \longrightarrow \sigma^*$ normally produce absorption in the ultraviolet–visible region. The other transitions require too much energy, so only molecules with $\pi$- or n-electrons give ultraviolet or visible spectra.

## 5.9 CHROMOPHORES

A compound with a carbon–carbon double bond has $\pi$ electrons and is a possible chromophore. An increase in the extent of delocalisation of electrons in a system containing double bonds increases absorption. Figure 5.9A shows this effect in the UV absorption spectra of the compounds $CH_3(CH{=}CH)_nCH_3$ when $n = 3$, $n = 4$ and $n = 5$. The wavelength of the absorption shifts to longer wavelengths, and the intensity of the absorption increases. $\beta$-Carotene, with eleven conjugated double bonds, absorbs at 450 nm and is therefore orange.

**FIGURE 5.9A**
The UV Absorption
Spectra of
$CH_3(CH{=}CH)_nCH_3$ for
(a) $n = 3$, (b) $n = 4$,
(c) $n = 5$

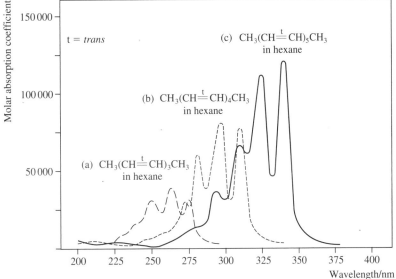

*An increase in the extent of
delocalisation of electrons
increases the absorption.*

**FIGURE 5.9B**
TheVisible–UV
Absorption Spectrum of
β-Carotene

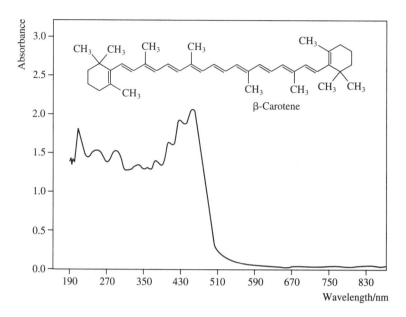

Most organic substances absorb in the UV and are therefore colourless. The
wavelength and the intensity of an absorption band can be changed by substituting
another group for hydrogen in the chromophore structure. The substituent need not
absorb by itself; it acts by shifting the energy of the electronic transition to a lower
level and therefore into the visible region of the spectrum. Such a substituent is called
an **auxochrome**. The benzene ring is a chromophore with maximum absorption at
204 nm in the ultraviolet. Unsaturated substituents which can form delocalised
systems with the benzene ring are auxochromes, moving the wavelength of maximum
absorption into the visible region of the spectrum. The ion of 4-nitrophenol has
maximum absorption at 400 nm and therefore appears yellow.

4-Nitrophenolate      $^-O$ —⬡— $NO_2$

The azo group is an effective auxochrome because it links two benzene rings by means
of a system of delocalised bonds.

⬡— N=N —⬡— OH

4-Hydroxyazobenzene (orange)

*The wavelength and
intensity of absorption are
changed when a hydrogen
atom is replaced by an
auxochrome, e.g. —NO₂
and —N=N— ...*

⬡— N=N —⬡— $N(CH_3)_2$

4-Dimethylaminoazobenzene (yellow)

$Na^+\ ^-O_3S$—⬡— N=N —⬡— $N(CH_3)_2$

Sodium salt of methyl orange

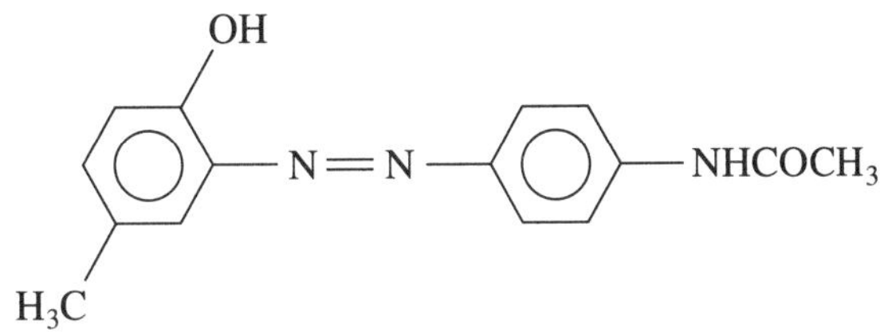

'Dispersol', Fast yellow G, a commercial dye

... —OH, —NH₂, and —SH.

*... —OH, —NH₂, and —SH.* Polar groups, e.g. —OH, —NH₂, —SH, can act as auxochromes by the use of non-bonded pairs of electrons. Benzene has an absorption peak at 255 nm [see Figure 5.1A], while phenylamine, $C_6H_5NH_2$ absorbs at 280 nm with six times the intensity. Azobenzene is colourless, while 4-$N,N$-diethylazobenzene, $C_6H_5—N=N—C_6H_4N(C_2H_5)_2$, is a yellow dye.

Some conjugated chromophores are:

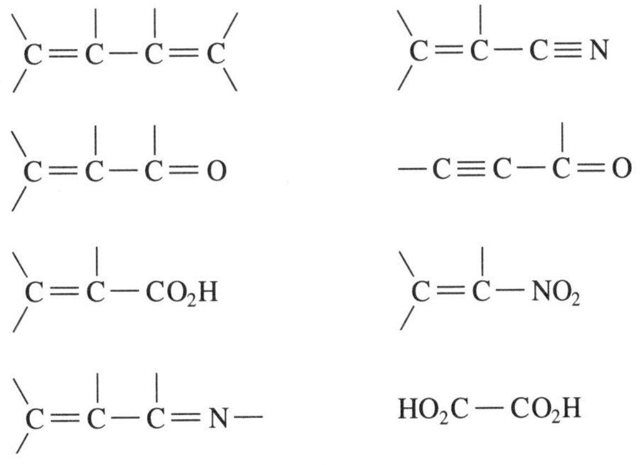

## CHECKPOINT 5.9

**1.** What types of molecular orbital are present in benzene?

**2.** Why is benzene colourless while nitrobenzene is yellow?

**3.** Many dyes contain an azo group. How does this group affect the absorption of visible light?

**4.** Predict whether the following compounds will absorb visible–UV light.

(a)

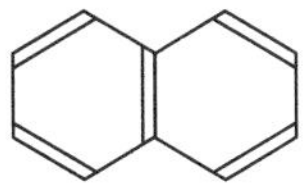

(b) $CH_3(CH=CH)_3CH_3$

(c) — Cl

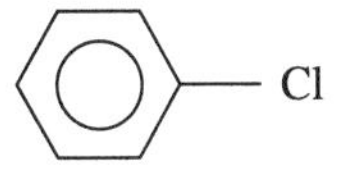

(d)

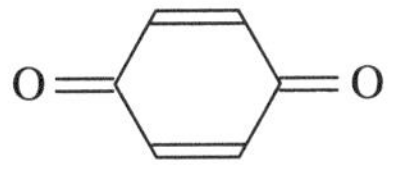

(e) $CH_3(CH_2)_5CH_3$

(f)

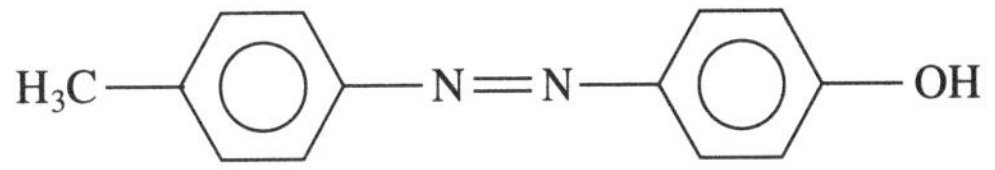

**5.** List the groups which cause a big shift in the wavelength of maximum absorption. Say what type of orbitals each of these groups possesses.

| Formula | $\lambda_{max}$/nm |
|---|---|
| $C_6H_6$ | 204 |
| $C_6H_5OCH_3$ | 217 |
| $C_6H_5CN$ | 224 |
| $C_6H_5NH_2$ | 230 |
| $C_6H_5CHO$ | 250 |
| $C_6H_5NO_2$ | 269 |

## 5.10 TRANSITION METAL COMPOUNDS

Transition metals are elements with an incomplete d-shell [see *ALC*, §24.1]. Many compounds of transition metals are coloured, both in the solid state and in solution. Table 5.10A gives examples.

| Ion | Electrons in d-subshell | Colour |
|-----|-------------------------|--------|
| $Ti^{3+}$ | $3d^1$ | Purple |
| $V^{3+}$ | $3d^2$ | Green |
| $Cr^{3+}$ | $3d^3$ | Violet |
| $Mn^{3+}$ | $3d^4$ | Violet |
| $Mn^{2+}$ | $3d^5$ | Pink |
| $Fe^{3+}$ | $3d^5$ | Yellow |
| $Fe^{2+}$ | $3d^6$ | Green |
| $Co^{2+}$ | $3d^7$ | Pink |
| $Ni^{2+}$ | $3d^8$ | Green |
| $Cu^{2+}$ | $3d^9$ | Blue-green |

*Transition metal compounds are coloured.*

**TABLE 5.10A**
The Colours of some Transition Metal Ions

*The colour is associated with incompletely filled d orbitals in transition metal atoms and ions, and the possession of unpaired electrons.*

We can seek an explanation for the colour in that important characteristic of transition metals: the partially filled d orbitals. When an atom or ion has partially filled d-orbitals, it possesses unpaired electrons. It is reasonable to suppose that the explanation for the colour of transition metal ions lies in these unpaired electrons. If an atom or ion has orbitals which differ in energy, then an electron can be promoted from an orbital of lower energy to an orbital of higher energy with the absorption of energy.

*In some complexes transition metal ions are surrounded by six ligands, which form coordinate bonds to the metal ion ...*

How can transition metal ions accomplish this change when the five d orbitals are **degenerate**? Orbitals of the same energy are described as degenerate. The explanation lies in the fact that transition metal ions are surrounded by six ligands which form coordinate bonds to the metal ion. A ligand may be a negative ion or a polar molecule with one or more lone pairs of electrons. Common ligands are water, $H_2O$, cyanide ion, $CN^-$, ammonia, $NH_3$, carbon monoxide, CO, 1,2-diaminoethane (en), $H_2NCH_2CH_2NH_2$, isocyanate, $NCS^-$ and edta, $(^-O_2CCH_2)_2NCH_2CH_2N(CH_2CO_2^-)_2$. The six ligands form an octahedral complex ion with the metal ion at its centre [see Figure 5.10A and *ALC*, §§ 24.13.6, 24.13.8, 24.13.9]. The colour of a transition metal complex depends on the metal, its oxidation state and the identity of any ligands involved. Table 5.10B illustrates this for cobalt complexes.

*... to give a complex ion of octahedral shape.*

**FIGURE 5.10A**
Octahedral Complexes of Cobalt

*The colour of the complex depends on the identity of the ligand.*

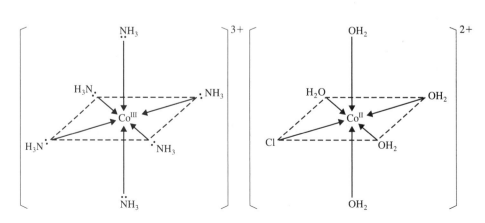

**TABLE 5.10B**
The Colours of some
Cobalt Complexes

| Complex | Colour of ion | Spectral colour absorbed | Wavelength/ (nm) |
|---------|---------------|--------------------------|-------------------|
| $[Co(NH_3)_6]^{3+}$ | Yellow | Indigo | 430 |
| $[Co(NH_3)_5H_2O]^{3+}$ | Red | Blue-green | 500 |
| $[Co(H_2O)_6]^{3+}$ | Pink | Blue-green | 500 |
| $[Co(NH_3)_5Cl]^{2+}$ | Purple | Green | 530 |
| $Co(edta)^-$ | Violet | Yellow | 560 |
| $[Co(en)_2Br(NCS)]^+$ | Blue | Orange | 610 |
| $[Co(en)_2Br_2]^+$ | Green | Red | >680 |

## 5.10.1 LIGAND FIELD THEORY

We need an explanation of the electron transitions in transition metal complexes. The **ligand field theory** looks at the interaction between ligands and the bonding orbitals of transition metal ions. In an isolated transition metal ion, the d orbitals are degenerate – all of the same energy. When there is interaction with ligands, however, the d orbitals are no longer degenerate. In the formation of a complex, ligands approach the central metal ion. This approach leads to repulsion between electrons in the ligand and electrons in the metal ion and therefore raises the potential energy.

*In a transition metal ion the five d orbitals are degenerate – of the same energy.*

**FIGURE 5.10B**
A View of the d Orbitals

*When six ligands coordinate to the ion they can approach either head-on to d orbitals or between d orbitals. The two approaches give rise to bonding d orbitals of different energies.*

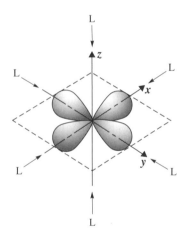

Orbital 1 has lobes along the x and y axes.

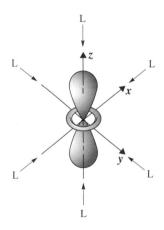

Orbital 2 has lobes along the z-axis.

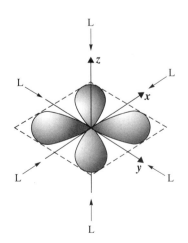

Orbital 3 has lobes in the angle between the x and y axes.

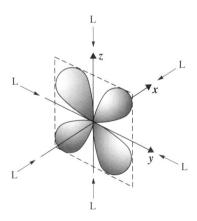

Orbital 4 has lobes between the y and z axes.

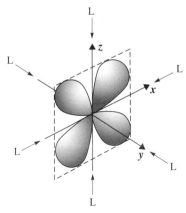

Orbital 5 has lobes between the x and z axes.

Consider the approach of ligands in the formation of an octahedral complex. If the electrostatic field were spherically symmetrical, the energy of all the d electrons would be increased by the same amount and the d orbitals would still be degenerate. However, the field is not spherically symmetrical. The five d orbitals occupy a three-dimensional configuration as shown in Figure 5.10B .

As the six ligands, L, approach along the x-, y- and z-axes, the approach to orbitals 1 and 2 in Figure 5.10B is head-on. The approach to orbitals 3, 4 and 5 in Figure 5.10B is not head-on. The ligands approach between the lobes of the d orbitals. The head-on approach raises the energy of orbitals 1 and 2 more than the sideways approach raises the energy of orbitals 3, 4 and 5. As a result, in the complex there are two d orbitals of higher energy than the other three. The difference in energy between them is the **ligand field splitting energy**, $\Delta E$. Energy can therefore be absorbed when an electron moves from a d orbital of lower energy into one of higher energy [see Figure 5.10C]. For most transition metals the energy difference $\Delta E$ is such that the light absorbed is in the visible part of the spectrum.

*The five d orbitals are divided into a group of two d orbitals which are of higher energy than the group of three d orbitals.*

**FIGURE 5.10C**
The Splitting of Five
d Orbitals into Two
Energy Levels

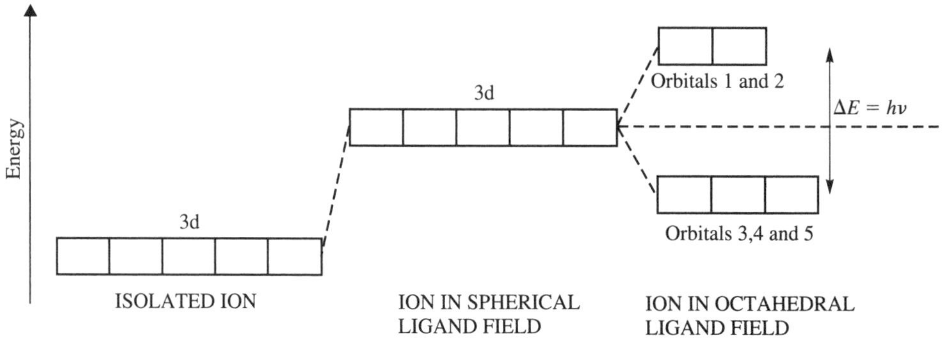

Energy

3d

Orbitals 1 and 2

$\Delta E = h\nu$

3d

Orbitals 3,4 and 5

ISOLATED ION

ION IN SPHERICAL
LIGAND FIELD

ION IN OCTAHEDRAL
LIGAND FIELD

*The difference between them is the ligand field splitting energy $\Delta E$. When electrons move from a d orbital to one of higher energy by absorbing $\Delta E$ the light absorbed is in the visible part of the spectrum.*

The splitting of d orbitals and the value of $\Delta E$ which determines the colour of a transition metal complex depend on a number of factors, including the following.

**1.** The identity of the metal and the number of d electrons present in a given oxidation state. Ions which have no d electrons, e.g. $Sc^{3+}$, and ions which have a $d^{10}$ configuration, e.g. $Cu^+$ and $Zn^{2+}$, have no possibility of a transition between d orbitals and are therefore colourless .

**2.** The number and nature of the ligands. Different ligands result in different values of $\Delta E$, the ligand field splitting energy. If aqueous ammonia is added to an aqueous solution of a copper(II) salt, the colour changes from pale blue to blue-violet. The ammonia ligands produce a bigger value of $\Delta E$ than do the water ligands in the original solution.

**3.** The shape of the complex [see Figure 5.10D].

**FIGURE 5.10D**
Colours of *Cis*- and *Trans*-
complex Ions of Cobalt

*The colour of a transition metal complex ion depends on the identity of the metal, the number and identity of the ligands and the shape of the complex.*

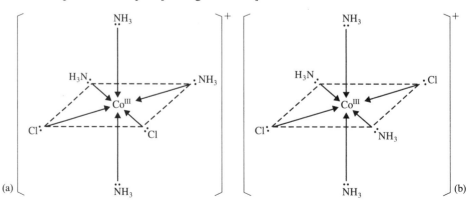

(a) *cis*-Tetraamminedichlorocobalt(III) ion (violet)     (b) *trans*-Tetraamminedichlorocobalt(III) ion (green)

The value of $\Delta E$ produced by a certain ligand varies from cation to cation. However, some ligands always produce a bigger change than others in all cations. Ligands can be ranged in order of the relative values of $\Delta E$ which they produce. This order is named the **spectrochemical series**.

Weak ligand field $\longrightarrow$ Strong ligand field

$$I^- < Br^- < Cl^- < F^- < OH^- < C_2O_4^{2-} < H_2O < NCS^- < NH_3 < py < en < NO_2^- < CN^- < CO$$

$\longrightarrow$ Increasing value of $\Delta E \longrightarrow$

(en = $H_2NCH_2CH_2NH_2$, py = porphyrin)

---

**CHECKPOINT 5.10**

1. How could you prove that the ligand $H_2O$ is essential to the blue colour of aqueous copper(II) ions?

2. Tests for $Fe^{3+}(aq)$ are the formation of (*a*) the blood-red thiocyanate complex, $Fe(CNS)^{2+}(aq)$ and (*b*) Prussian blue, $KFe[Fe(CN)_6](s)$. Give the approximate wavelength of light absorbed by (*a*) and (*b*).

3. When light is passed through a solution of $[Ti(H_2O)_6]^{3+}$ ions, light is absorbed. The absorption spectrum is shown in the figure.

(*a*) What colour is the light absorbed?

(*b*) What colour does the solution appear?

(*c*) If one quantum of light is absorbed, what happens to its energy?

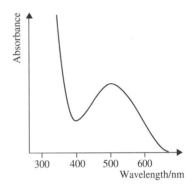

4. Zinc and copper are classified as transition metals. Why are the compounds of $Zn^{2+}$ and $Cu^+$ colourless?

---

## 5.11 APPLICATIONS OF VISIBLE-ULTRAVIOLET SPECTRA

### 5.11.1 IDENTIFICATION

*Compounds can be identified from the wavelength of maximum absorption and the molar absorption coefficient.*

The visible–UV spectrum can be used for the identification of compounds. The value of $\lambda_{max}$, the wavelength at which maximum absorption occurs, and the value of $\varepsilon$, the molar absorption coefficient in the solvent used, can be used to identify an unknown substance. Visible–UV spectra have broad absorption peaks, however, and identification is not a common application.

### 5.11.2 ANALYSIS

*The concentration of a solution can be found.*

Visible–UV spectrometry is widely used for finding the concentrations of solutions. The absorption at $\lambda_{max}$ is compared with a solution of known concentration.

### 5.11.3 INDICATORS

Most acid–base indicators are organic substances containing one or more chromophores. As the pH of the solution changes, the extent of electron delocalisation in the

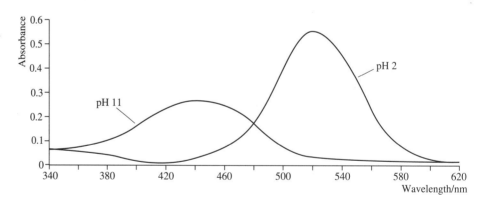

Methyl orange:          Yellow in alkaline solution                              Red in acidic solution

substance changes, and the colour changes. Methyl orange consists of yellow and red forms. Below pH 3, the indicator exists as the red 'acid' form, and above pH 10 it exists as the yellow 'basic' form.

The peak heights of the 'acidic' and 'basic' forms in a given solution can be determined by colorimetry. Comparison with the peak heights in pure solutions of each form gives the concentration of each form. This enables $K_a$ for the indicator to be found.

*Indicators have different absorption spectra in solutions of different pH.*

$$HIn(aq) \rightleftharpoons H^+(aq) + In^-(aq)$$

$$K_a = \frac{[H^+(aq)]\,[In^-(aq)]}{[HIn(aq)]}$$

If $K_a$ is known, the pH of the solution can be found.

**FIGURE 5.11A**
Absorption Spectrum of Methyl Orange

substance changes...

### 5.11.4   USING U–VISIBLE SPECTROMETRY WITH CHROMATOGRAPHY

The eluate from a chromatography column can be collected in fractions which are analysed by means of their visible–UV spectra. If a **single-beam spectrophotometer** is used, it is not necessary to collect the eluate in fractions. The eluate can be passed continuously through a cell to obtain its spectrum. Single beam spectrometers can measure a spectrum rapidly, in less than one second. They operate by passing the beam through the sample first and then switching it to pass through the reference cell. A computer interprets the two readings and plots the results as a spectrum. Then the spectrum can be compared with the database of a computer containing the spectra of a large number of compounds. See for example the testing of race horses for dope [§ 1.6.4].

*The eluate from a chromatography column can be analysed. A single-beam instrument responds rapidly and can be used for continuous monitoring.*

### 5.11.5  REACTION KINETICS

In a chemical reaction the visible–UV spectrum of the reactants is replaced by that of the products. If there is a good separation between the two spectra, the change may be used to measure the speed of the reaction. For example, bromine slowly forms a complex with 4-bromo-$N,N$-dimethylphenylamine.

*The progress of a reaction can be monitored by means of the change in the spectrum of the reactants or the products.*

$$N(CH_3)_2 + 2Br_2 \longrightarrow N(CH_3)_2 \cdot 2Br_2$$

The absorption at wavelength 400 nm can be measured and plotted against time to show the course of the reaction [see Figure 5.11B].

**FIGURE 5.11B**
The Visible–UV Absorption Spectra of Bromine and a Complex

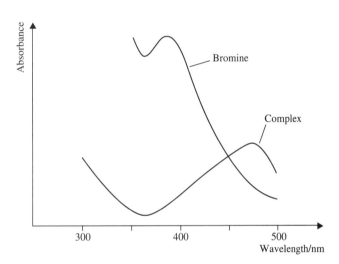

### 5.11.6  DETECTION OF IRON

Low concentrations of iron(II) can be measured by converting it into a coloured complex with 1,10-phenanthroline (phen) which absorbs strongly at 512 nm.

$$Fe^{2+}(aq) + 3phen(aq) \rightleftharpoons [Fe(phen)_3]^{2+}(aq)$$

Concentrations down to $1 \times 10^{-6}$ mol dm$^{-3}$ can be measured. The method can be used for iron(III) ions by reducing them to the iron(II) state. The iron content of a solid sample, e.g. a steel, can be found by reacting the sample with dilute sulphuric acid or dilute nitric acid and then reducing the iron salts formed to the iron(II) state which can be complexed with 1,10-phenanthroline.

*Iron and iron ions can be analysed by conversion into Fe$^{2+}$ ions which form a coloured complex with phenanthroline.*

1,10-Phenanthroline (phen)

### 5.11.7  ANALYSIS OF MANGANESE IN STEEL

In hot acidic solution, manganese(II) is oxidised by iodate(VII) to manganate(VII), which is intensely coloured.

$$2Mn^{2+}(aq) + 5IO_4^-(aq) + 3H_2O(l) \longrightarrow 2MnO_4^-(aq) + 5IO_3^-(aq) + 6H^+(aq)$$

The manganese content of a steel can be found by:

**1.** Reaction of manganese with dilute nitric acid.

**2.** Reduction of manganese ions in all oxidation states to manganese(II) ions by a solution of sodium sulphite.

**3.** Oxidation of manganese(II) to manganate(VII) by iodate(VII) (see equation above).

**4.** Addition of phosphoric acid to convert yellow iron(III) ions into a colourless complex.

*Manganese can be estimated by converting it into $MnO_4^-$ ions which are intensely coloured.*

**5.** Measuring the absorbance of $MnO_4^-$ ions. The percentage of Mn is read from a calibrated graph.

### 5.11.8  DETECTION AND ANALYSIS OF POLLUTANTS

Spectrophotometric methods are used to detect pollutants in water. The pollutants must be converted into a species which absorbs in the visible–UV region. For example,

- $NH_3 \longrightarrow H_2NHg_2I_3$, which is yellow
- $NO_3^-$ and $NO_2^- \longrightarrow$ orange-red azo dyes
- $SO_2 \longrightarrow$ red-violet dye
- $PO_4^{3-} \longrightarrow$ molybdenum blue
- phenols $\longrightarrow$ orange-red azo dyes
- $CN^- \longrightarrow$ blue dye
- $Cl_2$, As, B, $Br^-$, $F^-$, Se, $SiO_2$, $S^{2-}$, surfactants

Air pollutants which are converted into coloured compounds and analysed by visible–UV spectrophotometry include:

- $SO_2 \longrightarrow$ red dye
- $O_3 \longrightarrow$ oxidises KI to give $I_2$
- $NO_2 \longrightarrow$ azo dye

#### OTHER APPLICATIONS

*Other applications are mentioned.*

These include adding UV-absorbing inks to water marks in paper, postcoding valuables with UV-sensitive ink, using invisible but UV-fluorescent ink for signatures in building society savings books.

**CHECKPOINT 5.11**

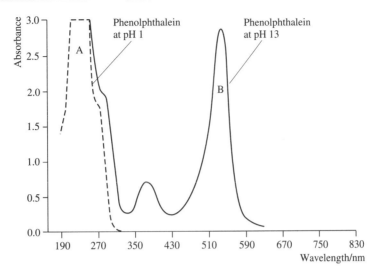

**1.** The visible–UV spectra of phenolphthalein at pH 1 (A) and pH 13 (B) are shown above.

(*a*) What range of wavelengths does phenolphthalein absorb in A and in B?

(*b*) What colours do these wavebands represent?

(*c*) What colour does phenolphthalein appear in A and in B?

(*d*) The formula of phenolphthalein is:

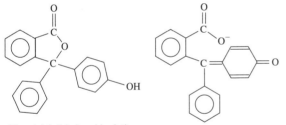

Phenolphthalein in acid solution

Phenolphthalein in alkaline solution

Explain how the difference in the spectra arises.

**2.** A sample containing 1.00 ppm nitrate ion was reduced, converted into a diazonium salt and then into an azo dye. The transmittance measured in a colorimeter was 35.5%. A water sample treated in the same way gave a transmittance of 45.2%. What is the concentration of nitrate ion in the water?

**3.** The $SO_2$ content of polluted air was converted into a dye and analysed by visible–UV spectrophotometry. A volume 10 $dm^3$ of air passed into the spectrophotometer gave an absorbance of 0.840. A sample containing 5.00 $\mu g$ $SO_2$ gave an absorbance of 0.700. What was the concentration of $SO_2$ in air in $\mu g\,m^{-3}$?

**4.** The manganese in a standard solution was oxidised to manganate(VII) and diluted to make a solution containing 1.00 mg Mn $dm^{-3}$. This solution had an absorbance of 0.316. A sample of wastewater, 10.00 $cm^3$, was oxidised to develop the colour of $MnO_4^-$ and diluted to 250 $cm^3$. The diluted solution had an absorbance of 0.296. What is the concentration of manganese in the wastewater?

## 5.12 FLUORESCENCE

*Some substances can absorb ultraviolet radiation ...*

If you attend discos, you may have noticed something odd happening to the white shirts which dancers are wearing. Under the disco lights, some white shirts glow with a blue colour. This transformation is caused by a phenomenon called **fluorescence**. It is a property of substances which absorb ultraviolet or visible radiation and re-emit visible light.

In order to fluoresce, a molecule first absorbs a photon. This increases its electronic energy and its vibrational energy. The increase in vibrational energy is small and is rapidly converted into kinetic energy by collisions with other molecules. The change in electronic energy is larger and it is re-emitted as electromagnetic radiation. The photon emitted has a lower energy than the photon absorbed and the emitted light

*... and convert part of its energy into kinetic energy ...*

therefore has a longer wavelength [Figure 5.12A]. Some molecules can therefore absorb ultraviolet radiation and emit visible radiation. The process is very rapid and emission stops as soon as the source of ultraviolet radiation is switched off.

**FIGURE 5.12A**
Fluorescence

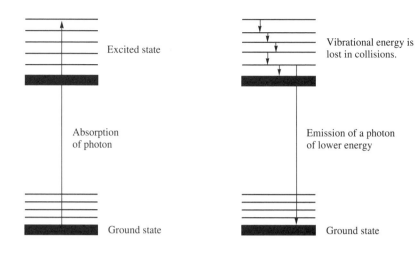

*... and emit visible radiation of lower energy. This is fluorescence.*

*Optical brighteners are fluorescent.*

Substances called **optical brighteners** are added to some laundry powders. They improve the appearance of white fabrics which have become tinged with yellow. The blue light emitted by the optical brighteners when they fluoresce masks the yellow tinge of the fabric. The fabric also appears brighter because more visible light is reaching the eye.

*While fluorescence occurs soon after excitation ...*
*... phosphorescence takes place over a longer time.*

Fluorescence is one form of **luminescence**: the ability of substances to absorb electromagnetic radiation or other forms of energy and emit visible light. If the return of the absorbing species from the excited state to the ground state occurs very soon after excitation, the phenomenon is called fluorescence. If the emission of light takes place over a longer time, the phenomenon is called **phosphorescence**. When chemical reactions provide the energy which is required to promote an electron, the luminescence is called **chemiluminescence**. It occurs in the reactions of luminol:

*Both are forms of luminescence ...*

Luminol

$$\text{NH}_2 \quad \text{O}$$

(structure of luminol showing a benzene ring fused to a ring containing NH₂, O, NH, NH, O groups)

*... which includes chemiluminescence and bioluminescence.*

When the reaction takes place in a living organism, the phenomenon is called **bioluminescence**. The firefly and the glow-worm emit light as a result of an enzyme-catalysed reaction between the compound called luciferin and oxygen.

Chemiluminescence is used in the detection and analysis of the pollutants ozone and nitrogen dioxide.

---

**CHECKPOINT 5.12**

**1.** Fluorescent substances absorb light of a certain wavelength and emit light of longer wavelength.

(*a*) Do the emitted photons possess more or less energy than the absorbed photons?

(*b*) What is responsible for the difference in energy?

**2.** Some substances continue to emit photons after the source of absorbed radiation has been removed. What is the name of this phenomenon?

**3.** What are (*a*) chemiluminescence, (*b*) bioluminescence?

**QUESTIONS ON CHAPTER 5**

**1.** The following compounds absorb light in the visible–UV region of the spectrum.

(a) $CH_3CH_2CHO$ (b) $CH_3CH_2NH_2$ (c) $CH_3CH=CH_2$

In each case say which electronic transitions are responsible for the absorption.

**2.** An aqueous solution of copper(II) sulphate is pale blue. When ammonia is added the colour changes to a dark blue. If 1,2-diaminoethane is now added to the ammoniacal solution, the colour changes to violet.

(a) Write the formulae of the copper(II) complex ions in the three solutions.

(b) Explain why the colour changes with different ligands.

**3.** Visible–UV radiation is absorbed by many organic substances.

(a) What is the name for the group in a molecule that absorbs visible–UV light?

(b) Give two examples of such groups

(c) State what characteristic such a group must possess.

(d) Explain how the groups absorb energy.

(e) What is an auxochrome? Give two examples.

**4.** Transition metal ions are coloured.

(a) Explain how the colour arises in e.g. $[Mn(H_2O)_6]^{2+}$

(b) Explain why the compounds of $Cu^+$ and $Zn^{2+}$ are colourless.

**5.** Benzene absorbs in the UV region of the spectrum, while nitrobenzene absorbs in the visible region. Explain why there is this difference.

**6.** A solution of the complex $FeSCN^{2+}$ of concentration $6.24 \times 10^{-5}$ mol dm$^{-3}$ has an absorbance at 580 nm of 0.437 in a 1.00 cm cell.

(a) Calculate the molar absorption coefficient at 580 nm of the complex.

(b) What percentage of the incident light is transmitted?

(c) A 2.50 cm$^3$ portion of a solution of iron(III) ions is treated with an excess of KSCN to form $FeSCN^{2+}$ and diluted to 50.0 cm$^3$. The solution gives an absorbance of 0.0625 at 580 nm in a 2.50 cm cell. What is the concentration of iron(III) in the original solution?

**7.** Electronic transitions occur when energy is absorbed from the visible or ultraviolet region of the electromagnetic spectrum.

(a) For each of the following compounds state one type of electronic transition that may occur: (i) $(CH_3)_2C=O$, (ii) $CH_2=CH—CH=CH_2$.

(b) Why does $CH_2=CH_2$ show an absorption band at a shorter wavelength than $CH_2=CH—CH=CH_2$?

(c) The hexaaquatitanium(III) ion, $[Ti(H_2O)_6]^{3+}$, is coloured. (i) How does the presence of the ligand affect the d orbitals of the uncomplexed titanium ion? (ii) How does the absorption of light affect the unpaired electron in $[Ti(H_2O)_6]^{3+}$ and hence give this hydrated ion a purple colour? [C]

**8.** Explain why aqueous ions containing transition metals are coloured, whereas aqueous ions of other metals are usually colourless. (You may wish to use $[Cr(H_2O)_6]^{3+}$ and $[Al(H_2O)_6]^{3+}$ as examples.)

(b) When an aqueous solution of the ligand L is mixed with an aqueous solution of a chromium salt, the following equilibrium is set up:

$$Cr^{3+}(aq) + 6L(aq) \rightleftharpoons [CrL_6]^{3+}(aq)$$

A similar equilibrium occurs with the ligand J, forming $[CrJ_6]^{3+}(aq)$.

The equilibrium constants for these reactions are both large and similar to each other, so that $[Cr^{3+}(aq)]$ in the solution is very small in the presence of the ligands.

Solutions **A**, **B** and **C** were made up by mixing 0.1 mol dm$^{-3}$ solutions of $Cr^{3+}$, L and J. The table below gives the volumes of each used.

|  | *Volumes of 0.1 mol dm$^{-3}$ solution/cm$^3$* | | |
|---|---|---|---|
| *Solution* | $Cr^{3+}(aq)$ | *L* | *J* |
| **A** | 2 | 98 | 0 |
| **B** | 2 | 0 | 98 |
| **C** | 2 | 49 | 49 |

The visible absorption spectra of the three solutions **A**, **B** and C are shown below:

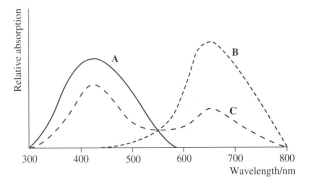

(i) What are the colours of solutions **A** and **B**?

(ii) The spectra show that the peak in the curve for solution **B** is at a longer wavelength than is the peak in the curve for solution **A**.

What deduction can be made from this fact about the size of the d orbital splitting in the two complexes?

(iii) The absorbance of a solution at a particular wavelength is proportional to the concentration of the ion responsible for the absorption.

Use this information and the given absorption spectra to suggest and explain which ligand, L or J, forms the stronger bond with $Cr^{3+}$. [C]

# 6

# INFRARED SPECTROMETRY

## 6.1 ORIGIN OF INFRARED ABSORPTION

*All organic compounds absorb in the infrared (IR) region of the electromagnetic spectrum.*

All organic compounds absorb in the infrared (IR). There are many more peaks in an IR spectrum than in a UV spectrum, making the IR spectrum better for identifying an organic compound. Books of recorded spectra are available for comparison. If the unknown compound is a new substance, its spectrum will not match with any recorded IR spectrum; yet it is still possible to infer a great deal about the structure of a compound. The C=O bond and the C—OH bond and others have characteristic absorption frequencies and can be identified. Functional groups can be identified from their IR spectra, and the IR spectrum is the most rapid method of detecting the series to which a compound belongs. Infrared spectrometry is valuable because of the great ease with which samples are prepared and spectra are taken.

*There are many peaks in an IR spectrum . . .*
*. . . and it is very helpful in identifying a compound.*

The operation of an IR spectrophotometer is similar to that of a visible–UV spectrophotometer. The source of radiation is a hot wire. The detector is a thermocouple (instead of a photomultiplier) which converts heat into an electric current. Since glass and quartz absorb IR radiation, the cells, mirrors and prisms are cut from large sodium chloride crystals.

*Different bonds give rise to different absorption peaks.*

Infrared radiation is part of the electromagnetic spectrum [see § 3.5]. The wavelength of infrared (IR) radiation is longer than that of visible and ultraviolet light. Absorption spectra in the visible–UV region arise from transitions between electronic levels in the molecules [§ 5.2]. The reason why absorption bands in the visible–UV spectra are broad is that the molecules are vibrating, and transitions occur between different vibrational modes within the same electronic level.

*The absorption arises from transitions between different vibrational modes within the same electronic level.*

If a molecule absorbs a photon of low energy, the energy may not be sufficient to promote an electron to a higher energy level, but it may change the molecule from one vibrational mode to another of higher energy [see Figure 5.1C]. The atoms in a molecule vibrate about their equilibrium positions even in the solid state. The frequency with which an atom vibrates depends on its mass and on the length and strength of the bonds it has formed. Bonds can absorb radiation of the same frequency as the natural frequency with which an atom vibrates ($1.20 \times 10^{13} - 1.20 \times 10^{14}$ Hz). This is in the IR region of the spectrum.

*An absorption band is identified by its wavelength $\lambda$ in nm . . .*
*. . . or by its frequency expressed as the wavenumber in $cm^{-1}$.*

The position of an absorption is measured by its wavelength $\lambda$ in nanometres, nm, or by its frequency expressed as the wave number $K$ in $cm^{-1}$. The infrared region extends from 4000 to 667 $cm^{-1}$. Absorptions that occur at the shorter wavelengths and higher frequencies are of higher energy than those at longer wavelengths. The wave number is directly proportional to the absorbed energy:

$$K = E/hc$$

and the wavelength is inversely proportional to the absorbed energy:

$$\lambda = hc/E$$

$$\lambda = 1/K$$

where $K$ = wave number, $\lambda$ = wavelength, $E$ = energy absorbed, $h$ = Planck's constant and $c$ = velocity of light [see Table 6.1A].

| Frequency, $v$/Hz | Wavelength, $\lambda$/m | Wavenumber, $K$/cm⁻¹ | Energy/ kJ mol⁻¹ |
|---|---|---|---|
| $1.20 \times 10^{13}$ | $2.50 \times 10^{-5}$ | 400 | $4.79 \times 10^3$ |
| $1.20 \times 10^{14}$ | $2.50 \times 10^{-6}$ | 4000 | $4.79 \times 10^4$ |

Infrared spectra originate from vibrations of bonded atoms. A molecule is not a rigid assembly of atoms. It can be thought of as a set of balls (the atoms) joined by springs of various lengths (the bonds). The springs can stretch, allowing the bonded atoms to move further apart or closer together while remaining in the same bond axis. The springs can bend so that the positions of the atoms change relative to the original bond axis. The various stretching and bending vibrations occur at certain quantised frequencies. When light of the same frequency as a particular vibration reaches a molecule, the molecule absorbs energy and increases the **amplitude** of that vibration. When the molecule returns from the excited state to the ground state, the absorbed energy is released as heat.

The frequency of infrared radiation must exactly match the frequency of the bond vibration. However, there is another condition for absorption in the infrared. The molecule must have a **dipole moment**, and for a particular vibration to result in the absorption of infrared energy the vibration must cause a change in the dipole moment of the molecule. Figure 6.1A shows the infrared spectrum of hydrogen chloride. Each line represents a transition from one combination of vibrational and rotational energy levels to another. The spacing between the lines is due to the moment of inertia of the molecule which gives it a resistance to rotational acceleration. This depends on the length of the H—Cl bond. The splitting of the lines into two is due to the presence of the two Cl isotopes. Chlorine-37, being heavier, gives H—$^{37}$Cl a greater moment of inertia: it needs more energy than H—$^{35}$Cl to force it through a change in vibrational or rotational motion.

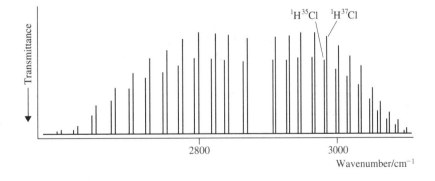

Figure 6.1B shows some vibrations of sulphur dioxide molecules. All the modes of vibration result in a change in the dipole moment of the molecule. Figure 6.1C shows comparable vibrations for carbon dioxide. As you see, the symmetrical stretch produces no change in dipole moment, therefore this stretch does not absorb infrared radiation. The infrared spectra are shown in Figures 6.1D and E.

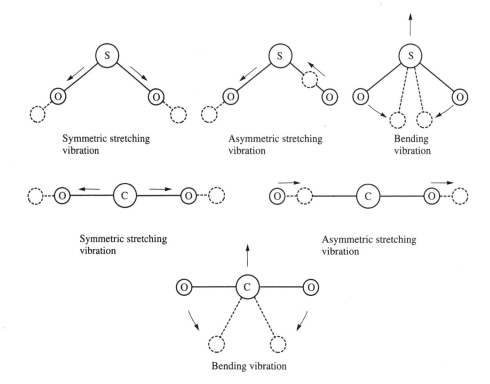

*The largest IR peaks arise
in O—H, C—O, C=O,
etc., where there is a large
change in polarity as a
result of vibration.*

Some IR absorption peaks are much stronger than others. The largest peaks arise
when there is a large change in polarity resulting from vibration of the bond. The
polar bonds O—H, C—O and C=O give rise to more intense absorptions than the
non-polar bonds C—H, C—C and C=C.

Bending vibrations normally require less energy and occur at lower wave numbers
than stretching vibrations. Stretching vibrations increase in order of bond strengths:
the triple bond is stronger than the double bond which is stronger than the single
bond. When the single bond joins the very small hydrogen atom, stretching vibrations
occur at low wavenumbers, e.g. O—H $3570\,\text{cm}^{-1}$ compared with O—D $2630\,\text{cm}^{-1}$.

*Bending vibrations require
less energy and occur at
lower wavelengths than
stretching vibrations.*

The spectra in Figures 6.1D and E are shown as percentage transmittance (the
percentage of the incident radiation that passes through the specimen) so that the
peaks extend from the top of the trace downwards. In molecules larger than $SO_2$ and
$CO_2$ it is not possible to identify each absorption. Each type of bond has a different
natural vibration frequency, but the same type of bond in different molecules is in a
different environment and its frequency changes; for example the C=O bond in
—CHO has a different absorption from the C=O bond in —CO₂H. There are no
two compounds with the same infrared spectrum.

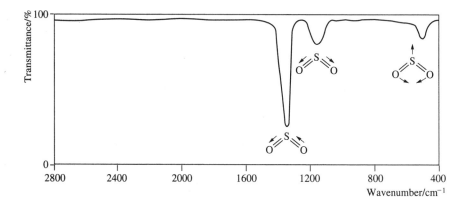

FIGURE 6.1E
The Infrared Spectrum of
Carbon Dioxide

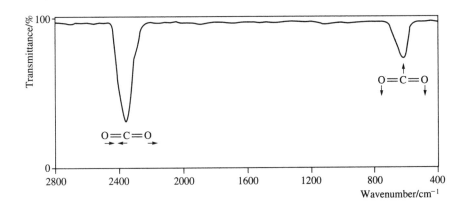

**FIGURE 6.1E**
The Infrared Spectrum of
Carbon Dioxide

*Raman spectra arise when transition to a different energy level occurs as a molecule scatters radiation.*

If a molecule scatters radiation and undergoes a transition to a different vibrational energy level, the emission is called **Raman emission**.

## 6.2 ORGANIC COMPOUNDS

Infrared spectroscopy is most frequently used in the identification of organic compounds. Organic compounds absorb infrared radiation over wavenumbers from $4000 \, cm^{-1}$ to $650 \, cm^{-1}$. All organic compounds absorb strongly in the infrared region of the spectrum. Many absorption bands are observed, in contrast to the the few absorption peaks observed in the ultraviolet region.

### 6.2.1 LOCALISED VIBRATIONS

A large molecule has a large number of vibrational modes. Some involve individual bonds or functional groups; these are called **localised vibrations**. Others involve the whole molecule. Localised vibrations are **stretching**, **bending**, **rocking**, **twisting** and **wagging**. The modes of vibration of the $CH_2$ group are shown in Figure 6.2A.

*Organic compounds show many IR absorption peaks. The modes of vibration of the $CH_2$ group are illustrated.*

Localised vibrations can be used to identify functional groups. Certain bonds always absorb in the same region, irrespective of the nature of the rest of the molecule, and this enables functional groups to be identified. The C=O bond absorbs in the region $1640–1815 \, cm^{-1}$ [see Table 6.2A]. The precise region depends on the rest of the molecule.

**FIGURE 6.2A**
The Modes of Vibration of the $CH_2$ Group (The '+' sign indicates motion from the page towards the reader, the '−' sign indicates motion away from the reader.)

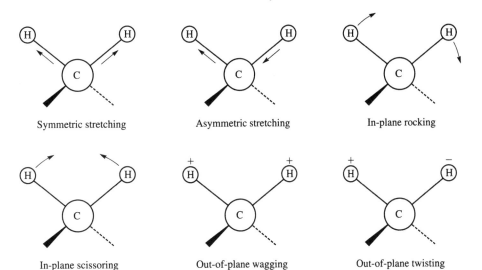

*Functional groups can be identified because they absorb in a certain region. . .*

**TABLE 6.2A**
Compounds Containing a C=O Group

| *Molecule* | *Wavenumber*/$cm^{-1}$ |
|---|---|
| Aliphatic aldehyde, RCHO | 1740–1720 |
| Aliphatic ketone, $R_2CO$ | 1725–1700 |
| Aromatic aldehyde, ArCHO | 1715–1695 |
| Aromatic ketone, $Ar_2CO$ | 1700–1680 |
| Aliphatic acid, $RCO_2H$ | 1725–1700 |
| Aliphatic ester, $RCO_2R'$ | 1750–1730 |
| Aliphatic amide, $RCONH_2$ | 1680–1640 |
| Aliphatic acid chloride, RCOCl | 1815–1790 |

The regions in which a number of groups absorb are shown in Figure 6.2B. Table 6.2B shows the wavenumbers of various groups. You need not try to remember them! You will always be able to refer to a table of wavenumbers when you need to identify a spectrum.

**FIGURE 6.2B**
The Infrared Absorption of a Number of Groups

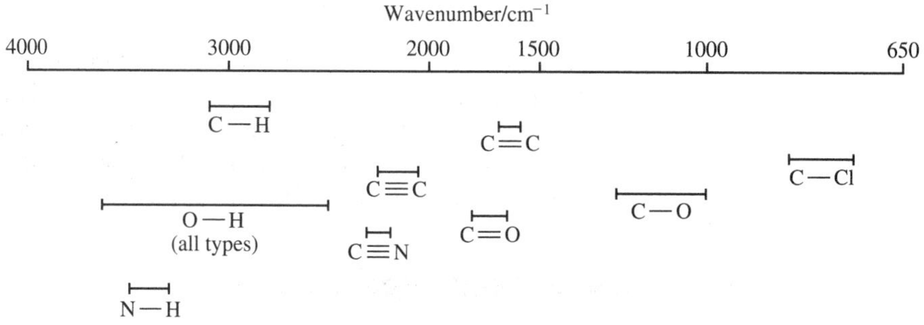


*. . . the precise wavelength depending on the rest of the molecule.*

*The wavenumbers of some functional groups are listed . . .*

*. . . for reference, not for memorisation.*

**TABLE 6.2B**
Wavenumbers of Functional Groups

| *Functional group* | | *Wavenumber*/$cm^{-1}$ |
|---|---|---|
| O—H | Aliphatic and aromatic | 3600–3000 |
| N—H | Primary, secondary, tertiary amines | 3600–3100 |
| N—H | Primary amines | 3500–3350 |
| C—H | Aromatic | 3150–3000 |
| C—H | Aliphatic | 3000–2850 |
| C≡N | Nitrile | 2280–2200 |
| C—H | of CHO | 2830–2720 |
| C—C | Arene | 1600 |
| O—H | Free | 3580–3670 |
| O—H | H-bonded in acids | 2500–3300 |
| O—H | H-bonded in alcohols and phenols | 3230–3550 |
| C≡C | Alkyne | 2250–2070 |
| $CO_2R$ | Ester | 1750–1700 |
| $CO_2H$ | Carboxylic acid | 1740–1670 |
| C=O | Aldehydes, ketones and esters | 1750–1680 |
| $CONH_2$ | Amides | 1720–1640 |
| C=C | Alkene | 1680–1610 |
| C—O—R | Aromatic | 1300–1180 |
| R—O—R | Aliphatic | 1160–1060 |
| C—Cl | | 800–700 |

### 6.2.2 FOUR REGIONS

*The IR spectrum can be considered as four regions . .*

The infrared spectrum can be considered as four regions:

4000–2500 cm$^{-1}$: single bonds between hydrogen and another element, e.g. C—H, O—H, N—H.

*. . . single bonds between hydrogen and another element . . .*

2500–2000 cm$^{-1}$: triple bonds, e.g. C≡C, C≡N.

*. . . triple bonds . .*
*. . . double bonds . . .*

2000–1500 cm$^{-1}$: double bonds, e.g. C=C, C=O, and also the N—H bending vibration.

*. . . and others.*

1500–400 cm$^{-1}$: other bond vibrations.

In general, the greater the bond strength, the higher the frequency of vibration.

#### HYDROGEN BONDING

*Hydrogen-bonded compounds show the O—H bond as a broad band.*

In hydrogen-bonded compounds, the O—H bond appears as broad bands instead of sharp peaks. This happens with alcohols and carboxylic acids.

Figure 6.2C shows the IR spectrum of Nujol, a hydrocarbon used as a mulling agent in the preparation of samples for IR spectroscopy [§ 6.4.4], which contains only C—C and C—H bonds.

**FIGURE 6.2C**
The IR Spectrum of Nujol Showing the Major Regions of the IR Spectrum

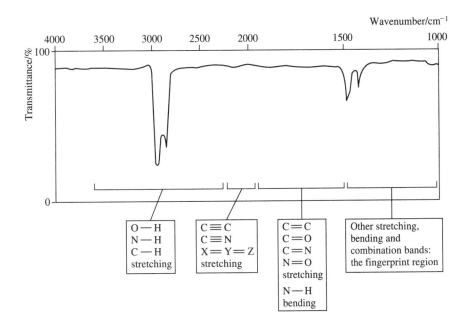

### 6.2.3 FINGERPRINT REGIONS

*In organic compounds many absorption bands arise below 1500 cm$^{-1}$. The exact form of these bands gives a 'fingerprint' region which can be used to identify an organic compound.*

The vibrations of the molecule as a whole give rise to a series of absorption bands at low energy, below 1500 cm$^{-1}$, with positions that are characteristic of that molecule [see the IR spectrum of benzene, Figure 6.6A]. Localised vibrations with frequencies below 1500 cm$^{-1}$ are therefore less useful for diagnostic purposes because of the overlap with molecular vibrations. Overtone bands arise, and combination bands can result from an interaction between two or more vibrations. The net result of all these vibrations is a region above 1500 cm$^{-1}$ where absorption bands can be assigned to functional groups and a region below 1500 cm$^{-1}$ containing many bands which are a **fingerprint** of the particular compound. Comparing the fingerprint region of a spectrum with that of an authentic sample of the compound is an extremely reliable method of identification – more reliable than a mixed melting point determination.

## 6.3  INSTRUMENTATION

*IR spectrometers employ prisms of rock salt or potassium bromide or calcium fluoride.*

The features of infrared spectrometers resemble those of visible–UV spectrometers. In simple systems, a prism resolves incident light into its spectrum; in more elaborate systems a prism and grating are used. Glass prisms cannot be used because they do not transmit infrared radiation: prisms of rock salt, potassium bromide and calcium fluoride are used.

*The source of radiation is often a hot nichrome wire . . .*

A spectrometer has a source of infrared light, e.g. a hot coil of nichrome wire, which emits radiation over the whole of the frequency range of the instrument. The beam is split into two of equal intensity. One beam passes through the sample to be examined, while the other is the reference beam [see Figure 6.3A]. The intensities of the two beams are compared after one has passed through the sample. The wavelength range over which the comparison is made is dispersed by means of a prism or a grating. The detector can be a photodiode (converting photons of radiation into small electrical currents) or a semiconductor (emitting electrons when radiation falls on it) or a photocell. The instrument scans the spectrum and plots absorption peaks against wavelength or frequency.

*. . . giving a beam which is split into two beams . . .*

**FIGURE 6.3A**
A Double-beam Infrared Spectrometer

*. . . one of which passes through the sample . . .*

*. . . while the other is the reference beam.*

*The intensities of the two beams are compared after one has passed through the sample.*

*The instrument plots percentage transmittance against wavelength or wavenumber.*

**1** Source of IR, a heated wire. The beam is split into two.

**2** One beam passes through the sample.

**6** The detector scans the frequencies. When the sample absorbs radiation, the sample beam is reduced in intensity, and the detector generates a signal.

**7** The recorder plots transmittance against wavenumber.

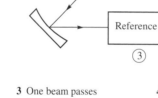

**3** One beam passes through the reference cell.

**4** The 'chopper' is a rotating disc with a segment cut out. It allows the two beams to pass alternately to the detector.

**5** The sodium chloride prism disperses the beam. Light of one frequency is focused on to the detector.

## 6.4  PREPARATION OF SAMPLES

*Samples are prepared for analysis . . .*

Compounds may be examined as solids, liquids or gases, or in solution.

### 6.4.1  GASES

*. . . as gases passed into a sodium chloride cell . . .*

The gas is passed into a cell about 10 cm long which is placed in the path of one of the IR beams. The end walls are made of sodium chloride which is transparent to IR radiation.

### 6.4.2  LIQUIDS

*. . . as liquids placed between discs of sodium chloride . . .*

A drop of the liquid is placed between discs of sodium chloride, made from single crystals. Sodium chloride is transparent to IR over 4000–625 cm$^{-1}$. For spectra below

$625\,cm^{-1}$ potassium bromide is used. The liquid must not dissolve sodium chloride or potassium bromide, and organic solvents must be free of water.

### 6.4.3  SOLUTIONS

*. . . as solutions to be placed between discs or used to fill a cell . . .*

The compound is dissolved in tetrachloromethane (carbon tetrachloride), $CCl_4$, or trichloromethane (chloroform), $CHCl_3$, to make a 1–5% solution. Solutions can be examined between discs, as for liquids, but if the absorption is low a longer path length is needed, and a cell of path length 0.1–1.0 mm is used. A similar cell containing solvent is placed in the path of the reference beam of the spectrometer. Intermolecular forces which operate in solids may well be absent in solutions. However, the solvent may alter the spectrum of the solute by hydrogen bonding. Solvents absorb in the infrared, and if the solvent absorbs over 65% of the incident radiation there is insufficient radiation to work the detector. A reason for the choice of tetrachloromethane and trichloromethane is that they do not absorb strongly in the wavelength region of interest.

### 6.4.4  SOLIDS

*. . . as solids, ground with a mulling agent to give stiff paste . . .*

*. . . which is pressed between discs of sodium chloride . .*

*. . . or ground with potassium bromide and compressed to form a disc.*

The solid, about 1 mg, can be finely ground with a drop of liquid hydrocarbon, e.g. Nujol or Kaydol, to produce a 'mull', a stiff paste. If C—H vibrations are to be studied a different mulling agent must be used, e.g. hexachlorobutadiene. The mull is pressed between discs of sodium chloride and inserted in the spectrometer. It may be necessary to avoid the bands due to the mulling agent. In this case, the solid can be ground with 10–100 times its bulk of pure potassium bromide, the mixture placed in a mould and compressed for some minutes at high pressure to form a transparent disc. The disc needs no other plates and is placed in the spectrometer. The reason for grinding the powder finely is that radiation is reflected off the surfaces of large particles. The number of resolved lines is often greater in the solid state spectra. Interactions with a solvent are absent, but intermolecular forces may complicate the spectrum.

## 6.5  FOURIER TRANSFORM INFRARED SPECTROMETRY (FTIR)

*A beam of light can be split into two beams . . .*

*. . . one made to take a longer path through the sample than the other. . .*

*. . . and the beams recombined to give an interference pattern . . .*

*. . . an interferogram . . .*

*. . . which can be converted by Fourier transformation into an absorption spectrum.*

A new method of taking an IR spectrum uses an interference technique. Radiation in the frequency range $5000–400\,cm^{-1}$ is split into two beams. One beam is made to traverse a longer path through the sample than the other. When the two beams are recombined an interference pattern is obtained. It is the sum of the interference patterns produced by all the wavelengths in the beam. The difference between the two paths is systematically changed, resulting in a change in the interference pattern with optical path difference. This pattern is called an **interferogram**. A computer in the instrument uses a mathmetical technique called **Fourier transformation** to convert the interferogram into an absorption spectrum. The advantages of FTIR are:

● It is not necessary to scan each wavelength separately, so the spectrum can be obtained in seconds.

● The speed makes it possible to scan the eluate of a chromatogram: the fractions do not have to be collected before analysis.

● High resolution is obtained because FTIR does not depend on a slit and a prism or grating.

● It can be used on very small samples: several scans can be added together if necessary.

● The mathematical technique makes it possible to subtract the spectrum of a pure compound from a mixture and reveal the spectrum of the other component or components in a mixture.

## 6.6  INTERPRETING IR SPECTRA

Benzene has a molecule with only C—C and C—H bonds. The IR spectrum [Figure 6.6A] shows not only the C—C arene bond at $1480\,\text{cm}^{-1}$ but also three C—H modes. These are the C—H stretching absorption at $3150–3050\,\text{cm}^{-1}$ and the C—H stretch in the plane of the ring at $1040\,\text{cm}^{-1}$ and the C—H stretch out of the plane of the ring at $680\,\text{cm}^{-1}$. Small peaks can be overtones or harmonics of fundamental vibrations.

**FIGURE 6.6A**
The IR Spectrum of
Benzene

*It is usually not possible to identify every peak in an IR spectrum . . .*

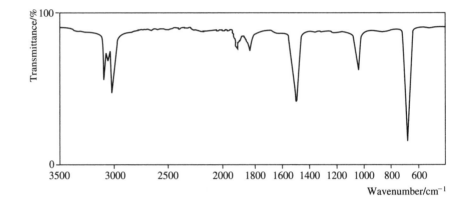

Many IR spectra are more complicated than that of benzene [Figure 6.6A]; it is not possible to identify every peak in an IR spectrum. Most organic compounds contain C—H bonds and analysis of C—H absorption does not provide much information. The best way to proceed is to identify important functional groups.

*. . . but many functional groups have peaks which are easy to identify.*

The functional groups C=O, N—H, C—O, O—H, C=C, C≡C and C≡N have peaks which are easy to recognise [see Table 6.2A and Figures 6.2B and C]. If a C=O group is detected over $1815–1640\,\text{cm}^{-1}$, the compound may belong to one of the series listed in Table 6.2A.

### 6.6.1  A KEY TO INTERPRETING SPECTRA

Say that C=O is present in a compound:

Is O—H also present? ⟶ Yes ⟶ indicates a carboxylic acid, —CO₂H.

No ⟶ Is N—H also present? ⟶ Yes ⟶ indicates an amide, —CONHR.

No ⟶ Is C—O also present? ⟶ Yes ⟶ indicates an ester, —CO₂R′.

No ⟶ Is the aldehyde C—H present? ⟶ Yes ⟶ indicates an aldehyde, —CHO.

No ⟶ Is the C=O peak a doublet? ⟶ Yes ⟶ indicates an acid anhydride,

(RCO)₂O.

No ⟶ indicates a ketone, RCOR′.

Say that C=O is absent in a compound:

Is C—O present? ⟶ Yes ⟶ indicates an ether ROR′.

No ⟶ Is N—H present? ⟶ Yes ⟶ indicates an amine, —NHR.

No ⟶ Is O—H present? ⟶ Yes ⟶ indicates an alcohol or phenol.

No ⟶ Is C=C present? ⟶ Yes ⟶ indicates an alkene or arene.

No ⟶ Is C≡C present? ⟶ Yes ⟶ indicates an alkyne.

No ⟶ Is C≡N present? ⟶ Yes ⟶ indicates a nitrile.

## CHECKPOINT 6.6

What can you deduce from the IR spectra in Questions 1–4?

**1.**

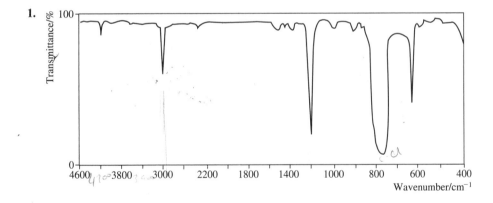

**2.**

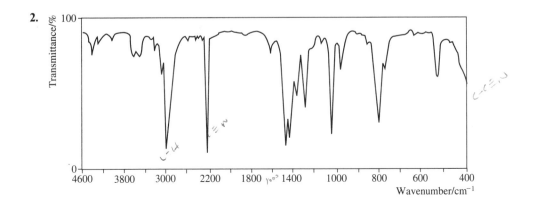

**3.**

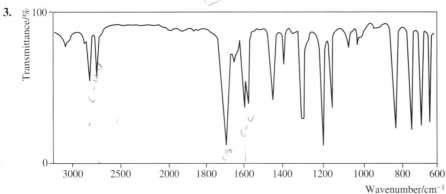

**4.**

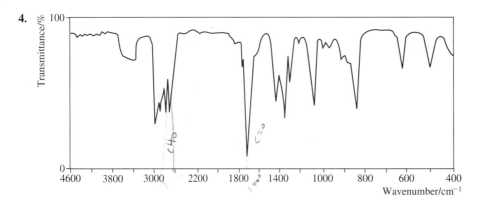

**5.** What type of compound gives rise to the following IR spectrum? How could you identify the compound?

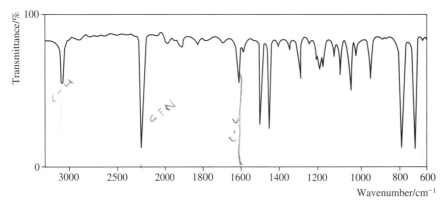

**6.** The IR spectrum of an organic liquid is shown below. What type of compound gave rise to the spectrum?

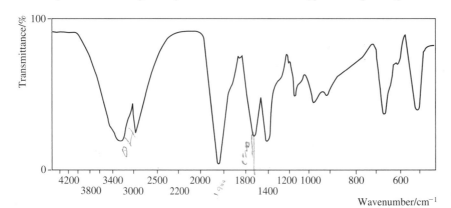

# 6.7  APPLICATIONS OF IR SPECTROMETRY

## 6.7.1  IDENTIFYING COMPOUNDS

*Applications of IR spectrometry include . . . . . . the identification of functional groups . . . . . and compounds . . .*

The presence of functional groups can be detected. In order to identify the compound, it may be necessary to study the mass spectrum or the nuclear magnetic resonance spectrum also. See § 8.9 and Questions on Chapter 8 for the use of **combined techniques**. When the identity of a compound is suspected, it may be readily confirmed by comparing the whole of the spectrum with that of an authentic specimen of the compound.

## 6.7.2  REACTION KINETICS

*. . . following the course of a reaction . . .*

The course of a reaction may be followed. When a secondary alcohol is oxidised to a ketone, the O—H band at $3570\,\text{cm}^{-1}$ is replaced by a C=O band at $1725\,\text{cm}^{-1}$. A portion of the reaction mixture can be withdrawn and examined to monitor the progress of the reaction.

## 6.7.3  FORENSIC SCIENCE

*. . . forensic science . . .*

IR spectrometry is used by forensic scientists. Part of the work of forensic scientists consists of providing evidence that a suspect was present at the scene of a crime or that a tool found in his possession, e.g. a screwdriver or a crowbar, was used to force an entry. Traces of paint and glass are frequently examined. A film of paint, consisting of pigments and a bonding resin, may have several layers of different colours. If paint fragments have been collected from a suspect's clothing or tools, they are first examined under a microscope. The differently coloured layers of paint may be very important in matching a chip of paint with painted woodwork at the scene of a crime or with a vehicle. The resins in a paint can be identified by IR spectrometry. Alternatively they can be pyrolysed and identified by gas chromatography coupled with mass spectrometry.

## 6.7.4  BREATH TEST

The police frequently have to decide whether a person has been driving while 'under the influence' of alcohol. The legal limit for drivers is 80 mg ethanol per $100\,\text{cm}^3$ of blood. Ethanol passes from the stomach into the blood stream and into the lungs. The ratio of ethanol in the blood to ethanol in exhaled breath has a constant value of 2300 : 1. The upper limit of ethanol in breath is therefore $335\,\mu\text{g}$ per $100\,\text{cm}^3$ of breath. If a driver is found to have more than 40 mg of ethanol per $100\,\text{cm}^3$ of exhaled breath, he or she is prosecuted.

*. . . 'breathalysing' drivers by measuring the absorption of IR by ethanol at $3340\,\text{cm}^{-1}$ . . .*

The police conduct a quick roadside test with a type of fuel cell. People can buy these to keep in readiness to breathalyse themselves when they feel that they may be over the limit. The fuel cell catalyses the oxidation of ethanol and generates an emf which is converted into a signal that indicates the level of ethanol in the breath. If the roadside test suggests that the driver is over the limit, he or she is taken to the police station where two breath samples are taken for accurate analysis.

The accurate method depends on the absorption of IR radiation by ethanol vapour [Figure 6.7A]. The IR spectrum shows absorption peaks at $3340\,\text{cm}^{-1}$ due to O—H and $2950\,\text{cm}^{-1}$ due to C—H. The O—H absorption cannot be used because water vapour in the breath affects the reading therefore the C—H absorption is used.

**FIGURE 6.7A**
The IR Spectrum of
Ethanol

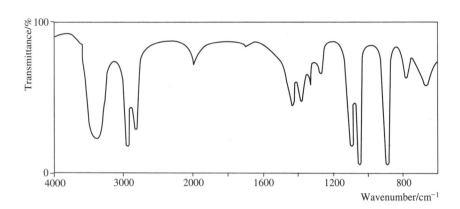

A commercial instrument has a beam of IR radiation from a heated filament which is split into two so that it can pass through the sample chamber and also a reference chamber. A filter ensures that only radiation of 2950 cm$^{-1}$ is measured. The sample container holds about 75 cm$^3$ of ethanol vapour. The instrument is calibrated to give a reading of breath ethanol concentration.

There are other substances that absorb at the same wavenumber. Butane is one. It is not normally present in the breath, but it may be after solvent abuse. It soon disappears from the breath, and if an interval of 20 minutes is allowed before testing butane does not appear. Propanone also absorbs at this wavenumber, and it is present in the breath of diabetics. This problem is solved by including in the analyser a semiconducting detector which is more sensitive to propanone than to ethanol. Then the ethanol measurement can be corrected for propanone concentration.

### 6.7.5  MONITORING AIR POLLUTION

Infrared spectrometry is used in the detection and analysis of air pollutants, in particular carbon monoxide. Air is passed through a long (100 cm) cell and the absorbance due to carbon monoxide is recorded. The Fourier transform method [§ 6.5] is used.

*... monitoring air pollution ...*

*... in combination with mass spectrometry and nuclear magnetic resonance.*

Laser spectroscopy is also used for detecting air pollutants. Examples of the use of atomic absorption spectrometry [§ 4.9.4], visible–ultraviolet spectrometry [§ 5.11.8], gas chromatography–mass spectrometry [§ 8.8.6] and high-performance liquid chromatography–mass spectrometry [§ 1.6.3] are mentioned elsewhere.

## QUESTIONS ON CHAPTER 6

What can you say about the IR spectra of the organic compounds shown below? It is not possible to identify each compound, but you will detect functional groups and be able to say whether the compound is aliphatic or aromatic.

**1.**

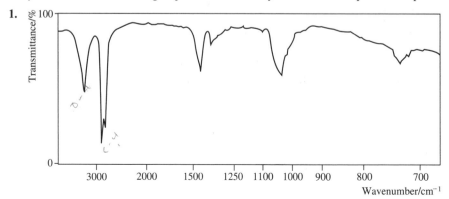

**2.** A liquid of $M_r$ 98.

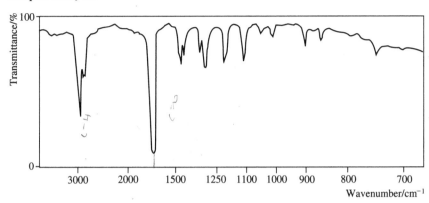

**3.**

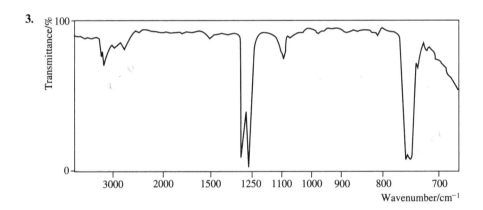

**4.**

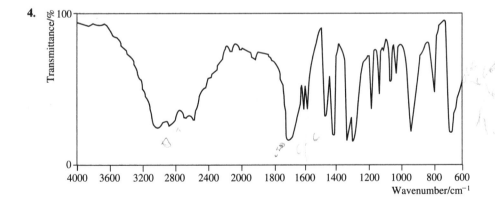

**5.**

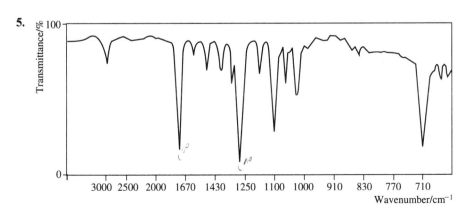

**6.** Identify the functional group in the compound which gave the following IR spectrum. Is the compound aliphatic or aromatic?

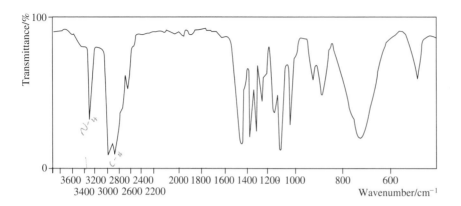

**7.** (*a*) Explain why infrared radiation is absorbed by the molecule HCl but not by the molecules $H_2$ and $Cl_2$.

(*b*) Explain what occurs in the HCl molecule when infrared radiation is absorbed.

(*c*) The simplified infrared spectrum below is that of an organic compound.

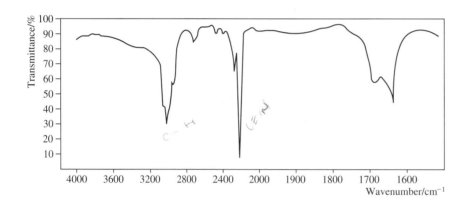

(i) Identify two main functional groups on the spectrum.
(ii) This compound has composition by mass: C, 67.9%; H, 5.7%; N, 26.4%, and $M_r$ of 53.
Suggest a structural formula for the compound.

[C]

**8.** The recycling of plastics could become cheaper and more efficient if the sorting of different types of plastic waste could be automated. One way of distinguishing between different types of plastic is by using infrared radiation. The plastics absorb radiation in different parts of the spectrum. The spectra **A**, **B** and **C** shown were produced by three common types of waste plastic.

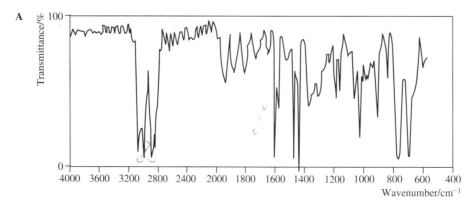

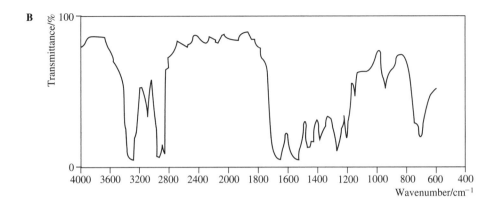

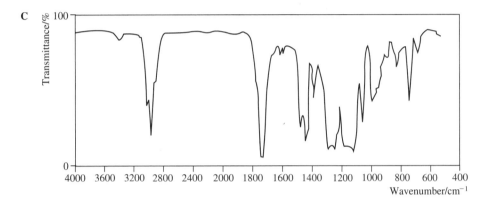

(*a*) Explain why plastics absorb radiation in the infrared region of the electromagnetic spectrum.

(*b*) Explain why the different plastics absorb in different regions of the infrared spectrum.

(*c*) For each plastic, suggest:
(i) an absorption which could be used to distinguish it,
(ii) its identity.

[C]

# 7

# NUCLEAR MAGNETIC RESONANCE SPECTROMETRY

## 7.1 BUCKMINSTERFULLERENE

*New allotropes of carbon have been discovered, and named fullerenes.*

In 1984, Harold Kroto and David Walton of Sussex University, UK, were engaged in space research. They were investigating the theory that large molecules containing up to 30 carbon atoms might be created in the space surrounding red giant stars. They directed a laser at a graphite target and detected the molecules they were looking for. They also detected by mass spectrometry molecules of relative molecular mass 720. They wondered whether these might be molecules of formula $C_{60}$, but they did not have enough material for a spectroscopic study.

*The structure of $C_{60}$ was worked out by spectroscopic methods of analysis . . .*

In 1980, Donald Hufmann in Arizona University, USA, and Wolfgang Kratschmer in Heidleberg University, Germany, had also made a strange form of carbon from heated graphite. In the light of the British discovery, they decided to have a more thorough look at it. By studying the mass spectrum, ultraviolet spectrum and infrared spectrum, they obtained results which supported the formula $C_{60}$. By X-ray diffraction and electron diffraction, they showed that the $C_{60}$ molecules are spherical and packed 1.04 nm apart.

Kroto and his team discussed possible structures for $C_{60}$. They tried to make a model sphere of 60 carbon atoms from hexagons of carbon atoms. They could not make a closed structure in this way, but when they tried a combination of 20 hexagons and 12 pentagons they obtained a closed structure resembling a football. They obtained electron microscope pictures which showed that there is in fact a mosaic of 20 hexagons and 12 pentagons on the surface of the molecule, which is 1 nm in diameter. The team decided to call the allotrope $C_{60}$ **buckminsterfullerene**, after the geodesic domes created by the architect Buckminster Fuller.

*. . . mass spectrometry, ultraviolet spectrometry, infrared spectrometry, X-ray diffraction, electron diffraction, electron microscopy . . .*

The final evidence for this new allotrope of carbon came from its **nuclear magnetic resonance (NMR)** spectrum. Kroto introduced $^{13}C$ into the molecule and measured the NMR spectrum. The spectrum was a single line! What this indicated, as you will appreciate a little further into this chapter, is that all 60 carbon atoms occupy identical positions in the molecule. This is true of the highly symmetrical structure of buckminsterfullerene.

*. . . and nuclear magnetic resonance spectrometry.*

There are other fullerenes [see Figure 7.1A]. In $C_{70}$, there are five different environments for carbon atoms, and the NMR spectrum of $^{13}C$-labelled $C_{70}$ has five NMR peaks.

**FIGURE 7.1A**
Fullerenes $C_{28}$, $C_{32}$, $C_{50}$,
$C_{60}$, $C_{70}$

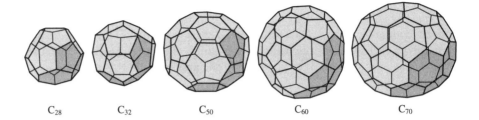

$C_{28}$      $C_{32}$      $C_{50}$      $C_{60}$      $C_{70}$

## 7.2 SPIN

*Atomic nuclei are charged and spin about an axis . . .*

*. . . giving rise to a magnetic moment.*

*The nuclear spin is quantised . . .*

*. . . therefore the magnetic moment of the nucleus is quantised.*
*A proton $^1H$ has spin quantum number $= \frac{1}{2}$.*

The basis of this powerful analytical technique, nuclear magnetic resonance, is that atomic nuclei can be thought of as tiny magnets. You are familiar with the idea that electrons possess spin [see *ALC*, §2.3.2]. Atomic nuclei also possess spin because they spin about an axis. The combination of charge and spin gives rise to a magnetic moment, that is, the nucleus behaves to some extent like a small bar magnet. The nuclear spin is quantised therefore the magnetic moment of the nucleus is quantised. For a proton, $^1H$, the spin quantum number $= \frac{1}{2}$. Other nuclei which contain an odd number of protons or neutrons or both also have a spin quantum number $= \frac{1}{2}$. Examples are $^{13}C$, $^{15}N$, $^{19}F$ and $^{31}P$. Two very common nuclei $^{12}C$ and $^{16}O$ have zero spin and zero magnetic moment and are invisible in NMR spectrometry. NMR is most widely used to identify $^1H$ nuclei. In such cases, it is often referred to as proton magnetic resonance, PMR In the case of buckminsterfullerene, $^{13}C$ was used.

The spin states of a certain nucleus have equal energy in the absence of a magnetic field. They are described as **degenerate** – of equal energy. If a magnetic field is applied, the spin states are no longer of equal energy. The magnetic moments of different nuclei may either align with or oppose the applied magnetic field [see Figure 7.2A]. The result is to split each spin state into two energy levels, $+\frac{1}{2}$ and $-\frac{1}{2}$ [see Figure 7.2B].

**FIGURE 7.2A**
Nuclear Spins

*NMR is most widely used to identify $^1H$ nuclei. When a magnetic field is applied, a nuclear spin can align with the field: spin $= +\frac{1}{2}$ . . . . . . or oppose it: spin $= -\frac{1}{2}$ . . .*

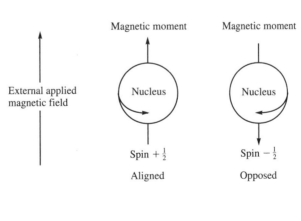

**FIGURE 7.2B**
The Splitting of Spin States in a Magnetic Field

*. . . therefore the spin state splits into two energy levels.*

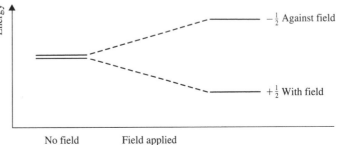

*Nuclei can be 'brought into resonance' by absorbing energy to move from the lower-energy spin state to the higher-energy spin state. The absorption of energy is called nuclear magnetic resonance. Energy absorbed = hv . . .*

*. . . v is in the radiofrequency region.*

Nuclei with spins that are aligned with the applied field can absorb energy and change their orientation from with the field to against the field. They are described as **flipping**. The absorption of energy that makes the nuclear spin **flip** is called **nuclear magnetic resonance**. The quantity of energy absorbed depends on the identity of the nucleus and the magnitude of the applied field. It is given by

$$E_{\text{absorbed}} = (E_{-1/2} - E_{+1/2}) = hv \qquad (h = \text{Planck's constant}; \S 3.6)$$

For typical applied magnetic fields, $v$ is in the radiofrequency region of the electromagnetic spectrum [see § 3.5]. It follows that the absorbed energy is much smaller than the energy changes associated with electronic, vibrational or rotational changes.

## 7.3  SHIELDING

*Protons in different parts of a molecule absorb energy at different applied fields.*

Protons in different parts of a molecule absorb energy at different applied fields. The reason is that small local magnetic fields are present in the neighbourhood of the protons. These magnetic fields originate because the applied field induces a circulation of the electron cloud round the nucleus, and this circulation produces a magnetic field in a direction which opposes the applied field. The nucleus experiences a reduced overall field and is described as **shielded** [see Figure 7.3A].

**FIGURE 7.3A**

Shielding of a Proton by its Electron Cloud

*The nucleus is shielded from the applied magnetic field . . .*

*. . . by a magnetic field due to its electron cloud. The extent of shielding depends on other atoms in the vicinity of the nucleus.*

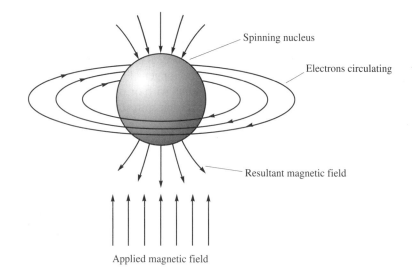

The extent of shielding depends on the electron density round the proton. In the —OH group in ethanol, the electronegative oxygen atom draws the electron cloud towards itself and the electron density round the proton is reduced, with a reduction in shielding. Protons in other parts of the molecule have a higher electron density round them and are shielded to a greater extent and require a higher applied field to bring them into resonance.

*The effect of electrons in π orbitals is to reinforce the applied magnetic field and deshield the protons.*

A different effect is found in molecules that contain electrons in π orbitals, especially when delocalisation is possible. The circulation of electrons that is induced by the applied field resembles a current in a loop of wire and generates a magnetic field. This field reinforces the applied field so that a lower applied field is needed to bring these protons into resonance. The protons are described as **deshielded** [see Figure 7.3B]. This effect is observed for protons in the benzene ring and in alkenes, alkynes and carbonyl compounds.

**FIGURE 7.3B**
Deshielding of an
Aldehyde Proton

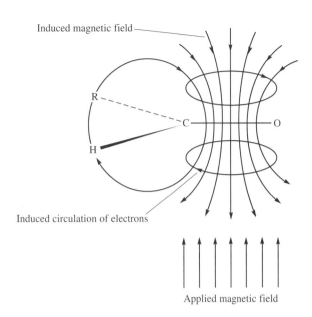

*The field required to bring
the protons in a substance
into resonance can be
measured.*

The field required to bring the protons in a substance into resonance can be measured.
The field depends on the frequency of the radiofrequency oscillator and on the
magnet employed; therefore a standard is needed. Tetramethylsilane (TMS),
$Si(CH_3)_4$, is chosen as a standard. It is chemically inert, miscible with most organic
solvents used and volatile (boiling temperature $27\,^{\circ}C$) so that sample materials can be
recovered from it. TMS gives a single NMR line because all the hydrogen atoms in it
are equivalent. The absorption line is at a higher frequency than other types of
organic protons.

*A standard substance
TMS, $Si(CH_3)_4$, is chosen
for reference. It gives a
single NMR line.*

The difference $\Delta B$ between the field required to bring the protons in a substance into
resonance and that required for TMS is measured. This difference in parts per million
of the applied field $B_0$ is called the **chemical shift**, $\delta$.

$\Delta B = $ *(Field required to
bring the protons in a
substance into
resonance) $-$ (Field
required to bring the
protons in TMS into
resonance)*
$\Delta B/B_0 = $ *Chemical shift
($B_0 = $ Applied field)*

$$\delta = \frac{\Delta B}{B_0} \times 10^6$$

where $\Delta B$ = difference in field, $B_0$ = applied field. For TMS, $\delta = 0$ by definition.

The value of $\delta$ for the protons in most common organic molecules is between 0 and
10. Highly shielded protons have low values of $\delta$, e.g. the protons in ethyne at $\delta = 2.4$.
Deshielded protons have high values of $\delta$, e.g. the protons in benzene at $\delta = 7.3$.

## 7.4 INSTRUMENTATION

The principle of an NMR spectrometer is illustrated in Figure 7.4A. The sample is
dissolved in a suitable solvent. Tetrachloromethane is often chosen as solvent because
many organic substances dissolve in it and because it has no protons which would
enable it to give rise to an NMR spectrum of its own. In some cases a solvent in which
the hydrogen atoms have been replaced by deuterium atoms is used. A small amount
of tetramethylsilane is added to serve as a reference compound (see below).

The frequency of the oscillation is fixed, usually 60 MHz or 100 MHz. The magnetic
field is varied until the exact condition for resonance is attained. Then the nucleus
absorbs energy as the nuclear spin 'flips' and aligns its magnetic moment against the
applied field.

**FIGURE 7.4A**
Features of an NMR
Spectrometer

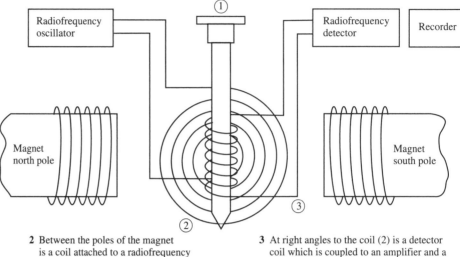

**1** A tube containing the sample is suspended between the poles of a permanent magnet.

**2** Between the poles of the magnet is a coil attached to a radiofrequency oscillator. This supplies enough energy for the nuclear spins to 'flip'.

**3** At right angles to the coil (2) is a detector coil which is coupled to an amplifier and a radio-recorder. When no energy is being absorbed by the sample, the detector picks up none of the energy from the oscillator coil. When the sample absorbs energy, a signal in the detector coil is recorded as a resonance signal.

*As the magnetic field is increased, successive protons come into resonance.*

As the electromagnetic field is increased, successive protons are brought into resonance. Signals can be detected from very small quantities of material, e.g. $0.5\,cm^3$ of a liquid or of a solution with a concentration down to $10^{-3}\,mol\,dm^{-3}$, a few milligrams of substance.

## 7.5  INTERPRETING SPECTRA

*The magnetic field needed to bring protons into resonance depends on the environment of the protons in the molecule . . .*

The NMR behaviour is a property of the nucleus so one might expect that protons in different environments, that is, different positions in the molecule, would have the same NMR spectra. In fact the low-resolution NMR spectrum of ethanol shows three absorption peaks at different values of applied field [Figure 7.5A].

**FIGURE 7.5A**
The NMR Spectrum of
Ethanol at Low
Resolution

*. . . e.g. in —$CH_2$— groups or —$CH_3$ groups or —OH groups.*

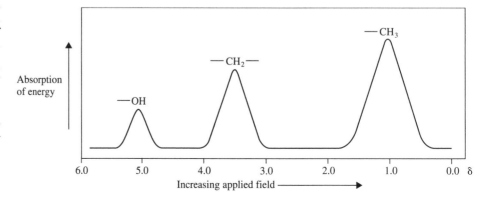

*The area of each peak is a measure of the number of protons in a certain environment . . .*

The areas under the peaks are in the ratio 1:2:3. The different values of field strength for the three peaks show that the single —OH proton, the two —CH$_2$— protons and the three —CH$_3$ protons all resonate at slightly different applied fields. Similar observations are made with other compounds, that is, peaks are produced at different field strengths by protons in different positions in the molecule, and the ratio of the areas of the peaks is the ratio of numbers of protons of each type.

NMR spectrometers superimpose an **integration trace** on the spectrum. The peak area is proportional to the height of the step on this trace [see Figure 7.5B].

**FIGURE 7.5B**
Low Resolution NMR
Spectrum of
Ethoxyethane

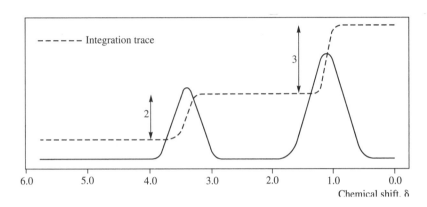

*. . . and is registered on the integration trace.*

The areas under the two peaks are compared by measuring the steps on the integration trace. In Figure 7.5B for ethoxyethane, C$_2$H$_5$OC$_2$H$_5$, the steps indicate that four of the protons are equivalent (two —CH$_2$— groups) and six protons are equivalent (two —CH$_3$ groups).

## 7.6  SPIN–SPIN SPLITTING

The chemical shift values and the integration trace supply information about the number and type of protons in a molecule of the substance. **Spin–spin** splitting gives further information. When the magnetic field is accurately controlled it is possible to obtain **high-resolution spectra**. (High resolution means the ability to distinguish between objects or lines that are close together.) In the high-resolution spectrum, some of the absorption peaks observed at low resolution are split into a number of components. In 1,1,2-trichloroethane,

*High-resolution spectra show a larger number of peaks due to spin–spin splitting.*

$$\begin{array}{ccc} & H_a & H_b \\ & | & | \\ Cl - & C - & C - Cl \\ & | & | \\ & Cl & H_b \end{array}$$

there are protons of two types, H$_a$ and H$_b$. There should be two resonance peaks, and the integration trace should show a ratio of 1:2 in height. The spectrum is shown in Figure 7.6A.

The spectrum contains five peaks: a triplet produced by H$_a$ and a doublet produced by H$_b$. The three peaks of the triplet are in the intensity ratio 1:2:1. The two peaks in

**FIGURE 7.6A**
A High-resolution NMR
Spectrum of 1,1,2-
Trichloroethane

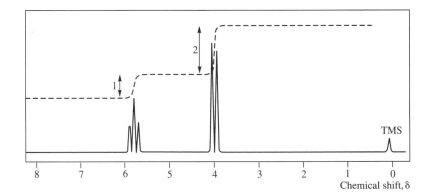

the doublet are equal in intensity. The splitting of the peaks can be explained. The applied magnetic field which the protons $H_b$ experience is modified by the local field which is produced by the magnetic moment of proton $H_a$. This local field may be aligned with or opposed to the applied field. In a group of 1,1,2-trichloroethane molecules, some have the local field of $H_a$ aligned with the applied field and others have it in opposition. Thus two slightly different energy states arise, and the peak is split into a **doublet**. In the case of the proton $H_a$, the applied field is modified by local fields produced by both $H_b$ protons. These local fields may both align with the applied field, or both may oppose it or one may align with the field and the other oppose it. In this way the peak becomes a **triplet** – 3 lines with intensities in the ratio $1:2:1$.

*The applied field which a certain proton experiences is modified depending on whether the field is aligned with or opposed to the local magnetic field of adjacent protons.*

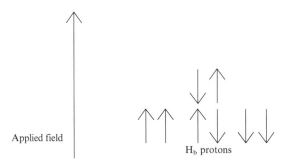

The splitting of peaks is governed by rules.

**1.** All chemically equivalent protons behave as a group.

**2.** A peak due to $n$ adjacent protons is split into $n + 1$ parts.

**3.** The intensity ratios of these **multiplets** is as shown in the triangle, which is called **Pascal's triangle**.

*If there are n protons adjacent to a certain proton, the peak due to that proton splits into n + 1 peaks of different intensities.*

| Number of chemically equivalent protons causing splitting | Relative intensities of lines in the splitting pattern |
|---|---|
| 1 | 1  1 |
| 2 | 1  2  1 |
| 3 | 1  3  3  1 |
| 4 | 1  4  6  4  1 |

The splitting pattern therefore gives information about the number of chemically equivalent protons adjacent to the proton under consideration. The high-resolution NMR spectrum of ethoxyethane is shown in Figure 7.6B.

**FIGURE 7.6B**
High-resolution NMR
Spectrum of
Ethoxyethane

*The splitting pattern gives information about the number of chemically equivalent protons adjacent to a certain proton.*

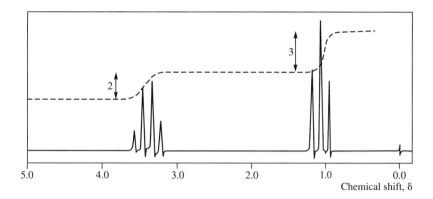

The —$CH_2$— and $CH_3$— groups within the molecule produce different absorptions. The protons in each —$CH_3$ group are chemically equivalent and they experience local modifications to the applied field due to the protons in each —$CH_2$— group. The result is a triplet of lines. The integration trace shows $\frac{3}{5}$ of the protons, confirming that this absorption is produced by —$CH_3$ protons. The two pairs of —$CH_2$— protons have the applied field modified by three adjacent protons on the —$CH_3$ groups. The result is a $1:3:3:1$ splitting pattern, as seen in Figure 7.6B. The integration trace shows $\frac{2}{5}$ of the protons, confirming that this absorption is produced by the —$CH_2$— protons.

## 7.7  IDENTIFYING LABILE PROTONS

The low-resolution NMR spectrum of ethanol is shown in Figure 7.5A. The high-resolution NMR spectrum shows splitting of the peaks.

*At high resolution, single peaks are resolved into multiple peaks.*

$$H_a-\overset{\displaystyle \overset{H_a}{|}}{\underset{\displaystyle \underset{H_a}{|}}{C}}-\overset{\displaystyle \overset{H_b}{|}}{\underset{\displaystyle \underset{H_b}{|}}{C}}-O-H_c$$

The $H_a$ protons have two adjacent $H_b$ protons and give a $1:2:1$ triplet. The $H_c$ proton has two adjacent $H_b$ protons and gives a $1:2:1$ triplet. The $H_b$ protons have three $H_a$ protons and one $H_c$ proton adjacent. The result is that the $H_b$ peaks are split by the $H_a$ protons into a $1:3:3:1$ quartet, and each peak of the quartet is split into a doublet by the $H_c$ proton aligning with or against the applied field; see Figure 7.7A.

**FIGURE 7.7A**
The High-resolution
Spectrum of Pure Ethanol

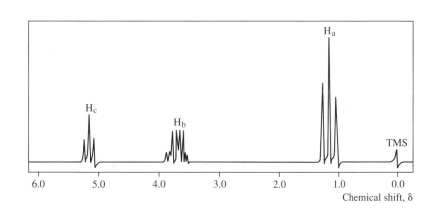

**FIGURE 7.7B**
The High-resolution
Spectrum of Impure
Ethanol

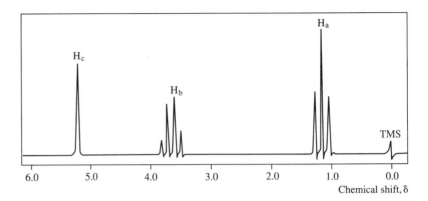

*Labile protons, e.g. in
—OH groups, can be
detected. When water is
present the high-resolution
spectrum changes.*

Figure 7.7A shows the high-resolution NMR spectrum of pure ethanol. When the sample contains traces of water, differences arise in the high-resolution NMR spectrum; see Figure 7.7B. The absorption of $H_c$ is a single peak, and that for $H_b$ is a 1:3:3:1 quartet. Some change must be taking place in $H_c$, the —OH proton. In the presence of water, proton exchange occurs:

$$C_2H_5OH' + HOH \rightleftharpoons C_2H_5OH + HOH'$$

This exchange occurs rapidly, and the $H_b$ protons experience the averaged effect of the spins of the protons which exchange with $H_c$ on the oxygen atom.

*When $^2H_2O$ is added, the
peak due to —OH
disappears.*

When $^2H_2O$ is added to the sample, the $H_c$ peaks at $\delta = 5.2$ disappear. The reason is that $H_c$ is replaced by $^2H$ (deuterium) which does not absorb in this region of the spectrum. The use of $^2H_2O$ to identify labile protons in a compound is an important technique in NMR spectroscopy [see Checkpoint 7.7].

Some proton chemical shift values, $\delta$ are tabulated. You do not need to memorise them as you will be able to refer to a table when you need to interpret spectra.

*Values of $\delta$, the chemical
shift, are given for
reference . . .
. . . not for memorisation.*

| Type of proton | Chemical shift, $\delta$/ppm |
|---|---|
| R—CH$_3$ | 0.85–0.95 |
| R—NH$_2$ | 1.0 |
| R—CH$_2$—R | 1.3 |
| R$_3$C—H | 2.0 |
| CH$_3$—CO$_2$R | 2.0 |
| R—COCH$_3$ | 2.1 |
| C$_6$H$_5$—CH$_3$ | 2.3 |
| R—C≡C—H | 2.6 |
| R—CH$_2$—Hal | 3.2–3.7 |
| R—O—CH$_3$ | 3.8 |
| R—O—CH$_2$R | 4.0 |
| R—O—H | 3.5–5.5 |
| RCH=CH$_2$ | 4.9–5.9 |
| C$_6$H$_5$—OH | 7.0 |
| C$_6$H$_5$—H | 7.3 |
| R—CHO | 9.7 |
| R—CO$_2$H | 11.0–11.7 |

**TABLE 7.7A**
Values of Chemical Shift,
$\delta$/ppm

**CHECKPOINT 7.7**

**1.** Identify the peaks in the NMR spectrum below, and suggest the identity of the compound which gives rise to these peaks. Explain why the peak at $\delta = 5.2$ is absent when $^2H_2O$ is present.

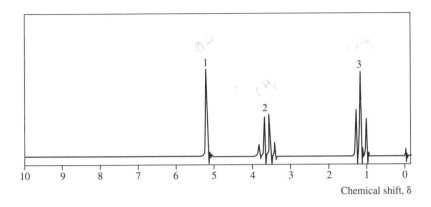

**2.** A compound contains the elements C, H, N. Suggest what groups may give rise to the peaks in its NMR spectrum. The peak at $\delta = 1.0$ is absent in $^2H_2O$. Explain how this happens. Suggest the identity of the compound.

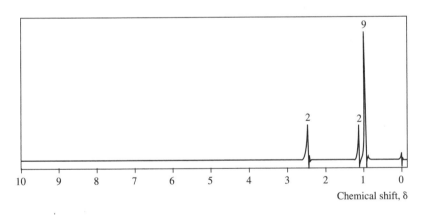

**3.** A compound contains the elements C, H, O. Identify the peaks in its NMR spectrum. Explain why the peak at $\delta = 11.7$ is absent when $^2H_2O$ is present.

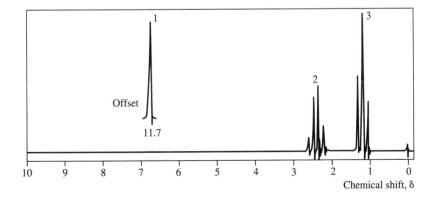

**4.** A compound is aromatic and contains only carbon and hydrogen. Identify the peaks in its NMR spectrum, and suggest what the compound may be.

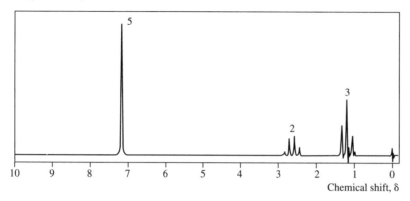

**5.** Identify the peaks in the NMR spectrum below, and suggest what the compound may be.

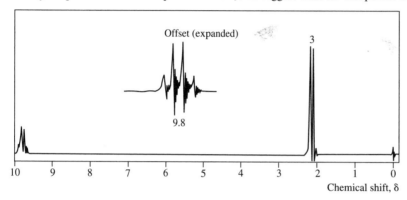

**6.** Identify the peaks in the NMR spectrum below and suggest what the compound may be.

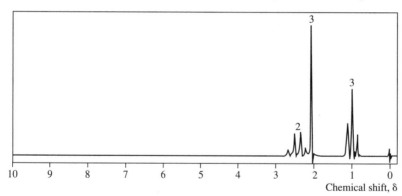

## 7.8 NMR IN MEDICINE

*Body scanners are NMR spectrometers . . .*

Many hospitals possess an instrument called a **body scanner**. This is an NMR spectrometer in which a patient can be placed so as to lie inside the large magnet. The patient has to remain still for about 20 minutes while scanning takes place. The machine obtains images called **magnetic resonance images** of soft tissue in any number of planes.

Protons in water, carbohydrates, proteins and lipids give different signals. The substances with a proton content sufficiently high to give a strong NMR signal are water and lipids. In different organs in the body the resonating protons have different

*...used to give an image of body tissues...*
*...which can assist in diagnosis.*

environments and therefore different signals. This enables different organs in the body to be differentiated in the magnetic resonance image.

There is no damage to the tissues and no known side effects, so patients can be scanned regularly. The technique has been used to diagnose cancer, multiple sclerosis, hydrocephalus (water on the brain) and other diseases.

## QUESTIONS ON CHAPTER 7

1. Outline the scope of NMR spectrometry in determining the molecular structures of organic compounds.

2. The NMR spectrum of an isomer of ethanol, is shown below.

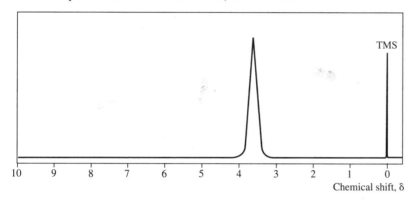

(a) Identify this compound and explain why its NMR spectrum is different from that of ethanol.

(b) Why is TMS used as a reference point?

3. Identify the ester with the following NMR spectrum.

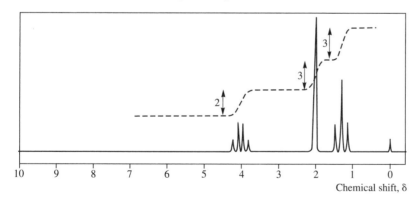

4. The NMR spectrum of a compound of formula $C_8H_{10}$ is shown below. Identify the compound.

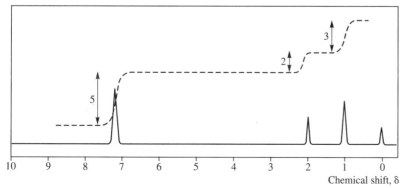

**5.** (*a*) In the NMR spectra of organic compounds, protons resonate at different chemical shift values ($\delta$). Explain why the chemical shift of protons in the methyl group of methylbenzene (2.3 $\delta$), is significantly lower than that of the hydroxy proton in phenol (7.0 $\delta$).

(*b*) The NMR spectrum shown below was obtained from a compound of formula $C_4H_{11}N$.

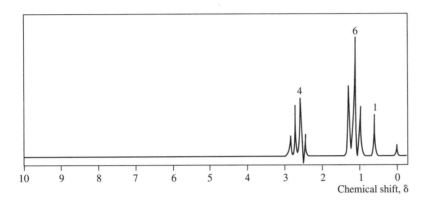

By referring to Table 7.7A for values of $\delta$, try to identify the compound, explaining how you arrive at your conclusion.

(*c*) Explain why, in medicine, NMR spectroscopy is particularly useful in such instruments as body scanners, compared with other spectroscopic techniques.

[C]

**6.** The following NMR spectra were produced by three isomeric compounds of formula $C_4H_{10}O$.

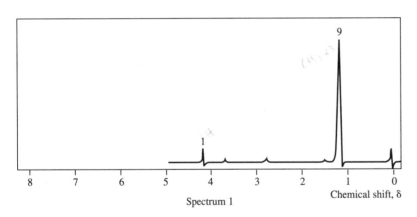

Spectrum 1

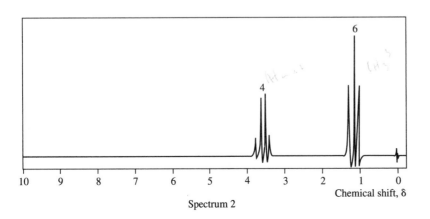

Spectrum 2

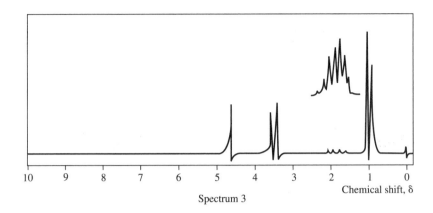

Spectrum 3

Chemical shift, δ

By studying each spectrum in turn,
(a) Describe the splitting pattern of each group of peaks.
(b) Identify the arrangement of hydrogen atoms responsible for each group.
(c) Draw the displayed (full structural) formula for each of the compounds.

[C]

# 8

# MASS SPECTROMETRY

## 8.1 HOW IT WORKS

The mass spectrometer is an instrument for determining atomic masses. It is based on the fact that when a stream of charged particles are sent through a magnetic field they are deflected from their path. In J. J. Thomson's historic apparatus, he sent a stream of electrons through a magnetic field and measured the deflection it experienced. From the deflection, he was able to calculate the ratio of charge to mass, $e/m$, of the electron [see *ALC*, §1.3 and Figure 1.4].

*When charged particles are sent through a magnetic field they change course . . .*

F. W. Aston developed this idea into the mass spectrometer. The deflection experienced by a particle depends on its mass, $m$, and its charge, $e$, and is determined

**FIGURE 8.1A**
A Mass Spectrometer:
How it works

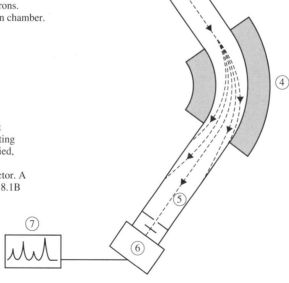

**2** The sample is injected as a gas into the ionisation chamber. Electrons collide with molecules of the sample and remove electrons to give positive ions. Some molecules break into fragments. The largest ion is the molecular ion.

**1** Heated filament gives electrons. They pass into the ionisation chamber.

**3** To this plate, a negative potential is applied (about 8000 V). The electric field accelerates the positive ions.

**4** An electromagnet produces a magnetic field. The field deflects the beam of ions into circular paths. Ions with a high ratio of mass/charge are deflected less than those with a low ratio of mass/charge.

**8** If the magnetic field is kept constant while the accelerating voltage is continuously varied, one species after another is deflected into the ion collector. A trace such as that in Figure 8.1B is obtained.

**5** These ions have the correct ratio of mass/charge to pass through the slit and arrive at the collector.

**7** Recorder. The electric current operates a pen which traces a peak on a recording.

**6** Amplifier. Here the charge received by the collector is turned into a sizeable electric current.

*... and the deflection is proportional to* m/e *(mass/charge). This is the basis of the mass spectrometer.*

by the ratio $m/e$. Massive particles of unit charge will be deflected less than light particles of unit charge. The higher the ratio of mass to charge, the smaller will be the deflection. A magnetic field will only deflect charged particles. Substances introduced into a mass spectrometer as atoms and molecules must first be ionised. This is done by bombarding them with a stream of electrons. Figure 8.1A shows how a mass spectrometer operates. Figure 8.1B shows the trace obtained for copper(II) nitrate.

**FIGURE 8.1B**
The Mass Spectrum of
Copper(II) Nitrate

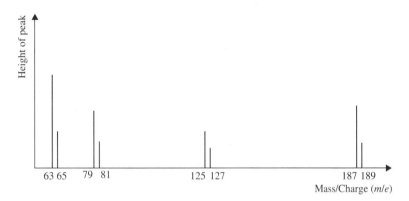

*Notes*

**1.** The height of each peak measures the relative abundance of the ion which gives rise to that peak.

**2.** The ratio of mass/charge for each species is found from the value of the accelerating voltage associated with a particular peak. Many ions have a charge of +1 **elementary charge unit**, and the ratio $m/e$ is numerically equal to $m$, the mass of the ion (1 elementary charge unit $= 1.60 \times 10^{-19}$ C).

**3.** The peaks on this trace correspond to the ions

$$63 = {}^{63}Cu^+,\ 65 = {}^{65}Cu^+,\ 79 = {}^{63}CuO^+,\ 81 = {}^{65}CuO^+,$$

$$125 = {}^{63}CuNO_3{}^+,\ 127 = {}^{65}CuNO_3{}^+,\ 187 = {}^{63}Cu(NO_3)_2{}^+,$$

$$189 = {}^{65}Cu(NO_3)_2{}^+$$

## 8.2  RELATIVE ATOMIC MASS

Figure 8.2A shows the mass spectrum of neon.

**FIGURE 8.2A**
The Mass Spectrum of
Neon

*Mass spectrometry is used to find relative atomic mass ...*
*... and isotopic masses ...*

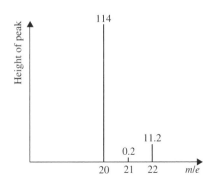

The average atomic mass of neon is calculated as follows.

Multiply the relative abundance (the height of the peak) by the mass number to find the total mass of each isotope present:

$$\text{Mass of } {}^{22}\text{Ne} = 11.2 \times 22.0 = 246.4 \, \text{u}$$

$$\text{Mass of } {}^{21}\text{Ne} = 0.2 \times 21.0 = 4.2 \, \text{u}$$

$$\text{Mass of } {}^{20}\text{Ne} = 114 \times 20.0 = 2280.0 \, \text{u}$$

$$\text{Totals} = 125.4 = 2530.6 \, \text{u}$$

$$\text{Average mass of Ne} = 2530.6/125.4 \, \text{u}$$

*... and the average relative atomic mass when an element has isotopes.*

$$= 20.18 \, \text{u}$$

The average atomic mass of neon is 20.2 u, and the relative atomic mass is 20.2.

## 8.3  RELATIVE MOLECULAR MASS

When a compound is ionised in the mass spectrometer, a molecular ion is formed by the loss of one electron.

*The ion with the highest value of* m/e *is the molecular ion.*

$$\text{M} \longrightarrow \text{M}^+ + \text{e}^-$$

The peak with the highest value of $m/e$ is due to the molecular ion. Its mass gives the molecular mass of the compound. This is not the only peak because in addition to generating molecular ions the beam of electrons ruptures chemical bonds with the formation of positively charged fragments. The largest peak corresponds to the most stable fragment. It is called the base peak and allocated a relative abundance of 100%. Other peak heights are expressed as percentages of the **base peak**.

**FIGURE 8.3A**
Mass Spectrum of
Methanol

*Other ions are fragments of molecules.*

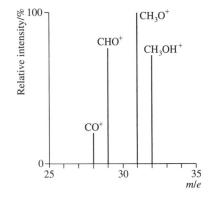

The mass spectrum of methanol is shown in Figure 8.3A. A peak for the molecular ion M is observed at $m/e = 32$. Peaks are obtained at $m/e = 31, 29$ and $28$. The peak at $m/e = 31$ can be assigned to the ion $CH_3O^+$ formed by the loss of one H atom. The loss of further H atoms gives the ions $CHO^+$ at $m/e = 29$ and $CO^+$ at $m/e = 28$.

*The largest peak is called the base peak ...*
*... and other peaks are measured against it.*

If isotopes are present, more than one molecular ion is formed [see § 8.5]. Some large molecules, e.g. polymers, are easily fragmented and do not give molecular ions.

## 8.4 FRAGMENTATION PATTERNS

A fast-moving electron has an energy of 10–70 eV (up to 7000 kJ mol$^{-1}$). In order to ionise an element an energy of 400–2400 kJ mol$^{-1}$ is required. In order to ionise an organic compound, an energy of 600–1000 kJ mol$^{-1}$ is required. The process

$$M \longrightarrow M^+ + e^-$$

takes place. $M^+$ has an unpaired electron: it is a radical. The energy of the bombarding electrons is so high compared with the bond energies that bonds in $M^+$ are broken, and fragments are formed.

*The molecular ion is fragmented by the beam of electrons. Charged fragments appear in the mass spectrum; uncharged fragments do not.*

The molecular ion carries a positive charge and has an unpaired electron. When the molecular ion $M^+$ fragments, only one fragment retains the positive charge. The other fragment is an electrically neutral species with an unpaired electron. Electrically neutral fragments do not appear as peaks in the spectrum. They include species such as the $\cdot CH_3$ radical and molecules such as CO and $H_2O$. The ions may fragment further. The way in which a molecular ion fragments depends on the relative stabilities of the species that can be formed. The ion $\cdot CH_3^+$ is less stable than the ion $R_3C \cdot^+$ in which the positive charge is less localised. The ease of formation of a number of fragment ions is:

$$CH_3^+ < RCH_2^+ < R_2CH^+ < R_3C^+ < CH_2{=}CH{-}CH_2^+ < C_6H_5CH_2^+$$

The products that are formed and show up in the mass spectrum depend on the energy of the ionising electrons.

Figure 8.4A shows the mass spectrum of butane. The peak at 58 is due to the molecular ion, $C_4H_{10}^+$. The peak at 15 corresponds to the $CH_3^+$ ion, that at 29 to the $C_2H_5^+$ ion and that at 43 to the $C_3H_7^+$ ion. The molecule has fragmented:

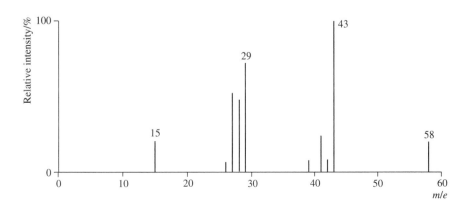

FIGURE 8.4B
The Mass Spectrum of
Ethanal

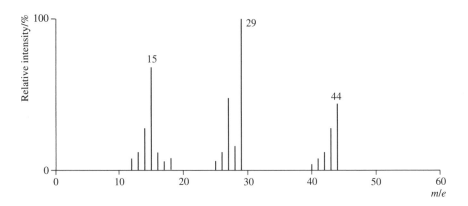

Figure 8.4B shows the mass spectrum of ethanal. The peak at 44 corresponds to the molecular ion. The peak at 15 corresponds to $CH_3^+$ and that at 29 to $CHO^+$. The molecule has fragmented:

FIGURE 8.4C
The Mass Spectrum of a
Hydrocarbon

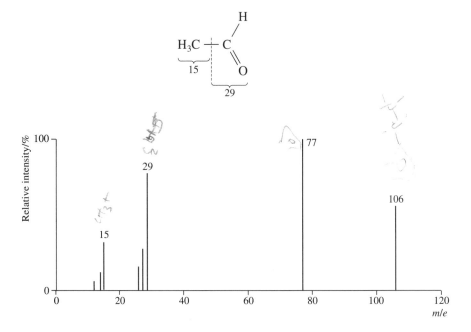

Figure 8.4C shows the mass spectrum of a hydrocarbon. How much can we deduce about its identity? The peak at 77 is very common in aromatic compounds; it corresponds to $C_6H_5^+$. The peak at 106 is probably the molecular ion. Subtracting 77 from 106 leaves 29, and a peak at 29 can be seen. This is likely to be due to $C_2H_5^+$. The peak at 15 corresponds to $CH_3^+$, which could be formed by the loss of $CH_2$ from $C_2H_5^+$. Putting together the groups $C_6H_5$, $C_2H_5$ and $CH_3$ gives $C_6H_5CH_2CH_3$, ethylbenzene, of relative molecular mass 106.

*The difference between the mass spectra of isomers is reviewed.*

Figure 8.4D shows the mass spectra of butan-1-ol and 2-methylpropan-2-ol. You can see that these isomers, $C_4H_{10}O$, break up in different ways. The mass spectrum of butan-1-ol shows a molecular peak at 74. The peak at 57 is due to $C_4H_9^+$, that at 43 to $C_3H_7^+$, 42 to $C_3H_6^+$, 41 to $C_3H_5^+$, 31 to $CH_2OH^+$, 29 to $C_2H_5^+$ and 27 to $C_2H_3^+$. The mass spectrum of 2-methylpropan-2-ol shows no molecular peak. The peak at 59 is due to the loss of $CH_3$ to give $(CH_3)_2COH^+$. The peak at 43 is due to $C_3H_7^+$ or $CH_3CO^+$, that at 41 to $C_3H_5^+$, 29 to $C_2H_5^+$ and 27 to $C_2H_3^+$.

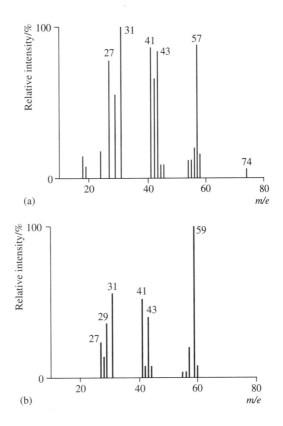
(a)

(b)

*An organic compound can be identified by comparing its fragmentation pattern with those of known compounds.*

The molecular ion and the fragmentation pattern can often identify an organic compound. However when a spectrum is compared with a library of spectra of known compounds for identification, the operating conditions, including the energy of the ionising electrons, must be quoted. In some cases electron impact ionisation does not give the required information because the molecule breaks up and no molecular ion is obtained. A different ionisation technique must be used.

*Some compounds fragment so readily that the molecular ion does not show in the mass spectrum.*

## 8.5   PEAKS AT $(M + 1)$, $(M + 2)$ AND $(M + 4)$

### 8.5.1   $(M + 1)$ PEAK

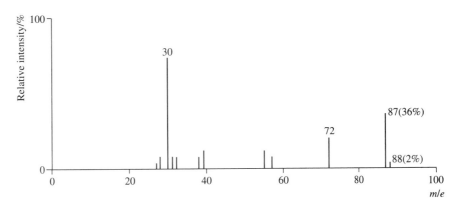

A mass spectrum can reveal the number of carbon atoms in a molecule of the substance. The mass spectrum shown in Figure 8.5A is that of a compound containing carbon, hydrogen and nitrogen only. As well as the molecular peak at $m/e = M$, there is a small peak at $m/e = (M + 1)$. This is due to the presence of $^{13}_{6}C$

atoms in the compound. Carbon-13 forms 1.1% of naturally occurring carbon. If $n$ is the number of carbon atoms in a molecule,

Height of $M$ peak/Height of $(M + 1)$ peak $= 100/1.1n$

and

Height of $(M + 1)$ peak/Height of $M$ peak $\times (100/1.1) = n$

In this spectrum, Height of $(M + 1)$ peak/Height of $M$ peak $= 2\%/36\%$

and

$n = (2/36) \times (100/1.1) = 5.05 = 5$ to the nearest whole number

*Peaks at* $(M + 1)$ *are due to the presence of* $^{13}_{6}C$ *and can be used to calculate the number of carbon atoms in the molecule.*

Establishing the number of carbon atoms in the molecule helps one to proceed further to explain other features of the spectrum. The molecular peak is at 87. With 5 carbon atoms in the molecule,

Molecular mass (87) $= 5 \times$ Relative atomic mass of carbon $(5 \times 12) + 27$

There is one N atom in the molecule; there cannot be two or there would be no mass units left for hydrogen. The formula must be $C_5H_{13}N$. The base peak at $m/e = 30$ could be $CH_2NH_2^+$ or $CH_4N^+$. There is a small peak at 72 corresponding to the loss of $CH_3$ from the molecule. Subtracting $CH_2NH_2$ from $C_5H_{13}N$ leaves $C_4H_9$. Combining $CH_2NH_2$ and $C_4H_9$ gives the formula $C_4H_9CH_2NH_2$. Considering the absence of a $C_2H_5$ peak at 29 or a $C_3H_7$ peak at 41, the formula must be $(CH_3)_3CCH_2NH_2$.

### 8.5.2  ($M$ + 2) PEAK

*Peaks at* $(M + 2)$ *with a ratio* $(M + 2)/M = \frac{1}{3}$ *are due to* $^{37}_{17}Cl$. *Peaks at* $(M + 2)$ *with a ratio* $(M + 2)/M = \frac{1}{1}$ *are due to* $^{81}_{35}Br$.

Some compounds show $(M + 2)$ peaks. They are usually caused by the presence of chlorine or bromine. In the case of chlorine, chlorine-37 is one-quarter of the total chlorine present as chlorine-35 and chlorine-37, and the ratio of Height of $(M + 2)$ peak/Height of $M$ peak $= \frac{1}{3}$. Bromine is a mixture of bromine-79 and bromine-81 in approximately equal amounts with bromine-79 being slightly more abundant. The presence of bromine gives an $(M + 2)$ peak with a ratio of Height of $(M + 2)$ peak/ Height of $M$ peak $= 1$.

### 8.5.3  ($M$ + 4) PEAK

A compound with two chlorine atoms or two bromine atoms has a peak at $(M + 4)$ [see Figure 8.5B].

**FIGURE 8.5B**
Mass Spectrum of
Dichloroethene, $C_2H_2Cl_2$

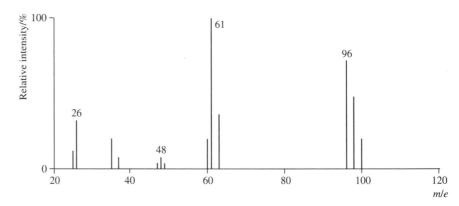

*Peaks at* $(M + 4)$ *are due to the presence of two* $^{37}_{17}Cl$ *or two* $^{81}_{35}Br$ *atoms in the molecule.*

The peak at 96 is the molecular peak for $C_2H_2{}^{35}Cl^{35}Cl^+$.

The peak at 98 is the molecular peak for $C_2H_2{}^{35}Cl^{37}Cl^+$.

The peak at 100 is the molecular peak for $C_2H_2{}^{37}Cl^{37}Cl^+$.

Peaks are seen at 35 and 37 in the ratio 3:1.

The peak at 61 is due to $C_2H_2{}^{35}Cl^+$.

The peak at 63 is due to $C_2H_2{}^{37}Cl^+$.

Again, the two peak heights are in the ratio 3:1.

The peak at 26 corresponds to $C_2H_2{}^+$.

## CHECKPOINT 8.5

**1.** A compound is known to be either $C_4H_{10}O_4$ or $C_7H_6O_2$. Both these formulae have $m/e = 122$. On a mass spectrum of this compound the heights of the peaks at $M$ and $(M + 1)$ are 15.0 and 1.15 respectively. Which is the formula of the compound?

**2.** Identify the compound which has the mass spectrum shown below.

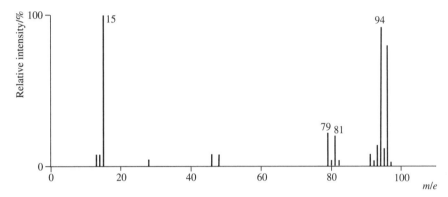

**3.** Interpret the mass spectrum shown below.

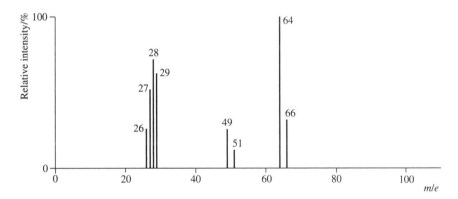

**4.** Mass spectrometry is used in conjunction with gas chromatography to detect and analyse pollutants in water. 2,4-Dichlorophenoxyethanoic acid (2,4-D) is a herbicide which can become a water pollutant. It has a distinctive spectrum which quickly reveals the presence of 2,4-D in water.

$$OCH_2CO_2H$$

Cl

Cl     2,4-D

The mass spectrum of 2,4-D shows distinctive peaks at $m/e = 162$, 175 and 220, with accompanying peaks at 164, 177 and 222. Which ions give rise to these peaks?

## 8.6  INSTRUMENTATION

### 8.6.1  INTRODUCING THE SAMPLE

*Mass spectrometry is very sensitive and requires only a tiny amount of sample ...*

Mass spectrometry is a thousand times more sensitive than infrared or nuclear magnetic resonance spectroscopy. One microgram or less of sample is sufficient to yield a mass spectrum. Samples must be introduced into the vapour phase. Gaseous samples are allowed to diffuse into the spectrometer. Volatile liquids are injected to vaporise at the low pressure inside the spectrometer. Involatile liquids and solids are placed on a ceramic tip or in a glass capillary or in a metal crucible. This is inserted into the spectrometer and heated to 300 °C so that the vapour will diffuse into the ion production region. A solution can be introduced by putting it into a crucible and allowing the solvent to evaporate. At the temperature of the ionisation chamber, 150–250 °C, the vaporised sample remains in the gas phase. At this temperature range, non-polar substances with relative molecular masses up to 1000 and polar substances with relative molecular masses up to 500 can be vaporised.

*... which is introduced into the instrument in the vapour phase ...*

### 8.6.2  METHOD OF PRODUCING IONS

Ions can be produced by the method of **electron impact** [see Figure 8.6A]. The electrons have energy 10–70 eV (1 eV = 1 electron volt = 96.5 kJ mol$^{-1}$). When an electron with a high energy collides with a molecule of the vaporised sample it displaces an electron from an outer electron orbital.

*... and ionised by electron impact.*

$$X(g) \quad + \quad e^- \longrightarrow X^+(g) \quad + \quad e^- \quad + \quad e^-$$

Molecule of sample    Electron of high energy    Ion    Electron displaced from X    Electron with reduced energy

Some doubly charged ions are formed. The value of $m/e$ gives a value for the mass which is half that of the real mass of the ion. Some negative ions are formed. They are not attracted by the magnetic field of the mass spectrometer.

**FIGURE 8.6A**
Electron Impact

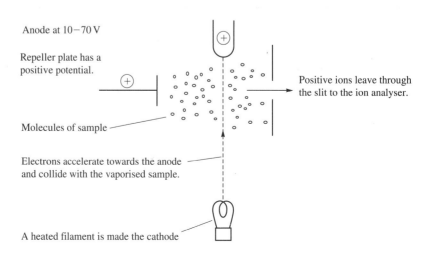

Anode at 10−70 V

Repeller plate has a positive potential.

Positive ions leave through the slit to the ion analyser.

Molecules of sample

Electrons accelerate towards the anode and collide with the vaporised sample.

A heated filament is made the cathode

## 8.7 HIGH RESOLUTION

*High-resolution mass spectrometry gives relative molecular mass to an accuracy of 1 in $10^8$ . . .*
*. . . and enables one to distinguish between isomers on the basis of accurate relative molecular mass.*

Mass spectrometry can be used to find the relative molecular mass of a compound to a high degree of accuracy. Where low-resolution mass spectrometry gives a value of $m/e = 101$, high-resolution mass spectrometry can give a value of $101.023\,456$ – an accuracy of 1 in $10^8$. The increase in resolution is obtained by means of a double-focusing mass spectrometer. This includes an electric sector before the magnetic field. Ions enter the electric sector and are focused into a narrower range of kinetic energy. When the ions enter the magnetic field they have a smaller range of energies, and the magnetic field is able to separate them with a higher resolution. For example, the compounds listed in Checkpoint 8.7 have the same value of $m/e$ at low resolution. The values of $m/e$ at high resolution are listed, and, as you will see, these make it possible to distinguish between the compounds.

━━━━━━━━━━━━━━━━━━ CHECKPOINT 8.7 ━━━━━━━━━━━━━━━━━━

Some accurate relative atomic masses are:
$A_r(C) = 12.000\,000\,0$,
$A_r(H) = 1.007\,824\,6$,
$A_r(N) = 14.003\,073\,8$,
$A_r(O) = 15.994\,914\,1$.

1. An organic compound contains C, H, N and O. The formulae suggested are $C_6H_4NO_2$, $C_6H_6N_2O$ and $C_7H_8NO$.

The mass spectrum gives a molecular ion of $m/e = 122.060\,585$. Which formula is correct?

2. A chemist is trying to decide whether a compound he has made has the formula $C_{10}H_{18}N_4$ or $C_8H_6N_2O_4$ or $C_{10}H_{10}O_4$. By high-resolution mass spectrometry he finds a relative molecular mass of 194.0328. Which formula is correct?

## 8.8 APPLICATIONS OF MASS SPECTROMETRY

### 8.8.1 DETERMINATION OF ATOMIC MASSES AND ISOTOPIC MASSES

*Applications of mass spectrometry include . . .*
*. . . determination of relative atomic masses and relative isotopic masses . . .*

This was covered in §8.2.

### 8.8.2 IDENTIFICATION OF COMPOUNDS

*. . . identification of compounds by their fragmentation patterns . . .*
*. . . and by means of accurate relative molecular masses . . .*
*. . . identifying components of foodstuffs . . .*

Mass spectrometry is used in combination with gas chromatography and high performance liquid chromatography [see §§1.5 and 1.6] to give rapid analyses of complex mixtures. The output from the mass spectrometer can be fed into a computer and the fragmentation pattern compared with the library of compounds in the database of the computer in order to identify an unknown compound. The accurate relative molecular mass is also used, as described in §8.7.

An application of this aspect of mass spectrometry is its use in checking for contaminants in foodstuffs.

### 8.8.3 FORENSIC SCIENCE

The sensitivity of the mass spectrometer makes it a valuable instrument for forensic scientists. The size of sample which they receive for analysis is often very small. A

... identifying drugs,
fibres, etc. in forensic
science ...

mass spectrum can be obtained on as little as $10^{-12}$ g. Small amounts of drugs can be identified. A fibre left at the scene of a crime can be compared by mass spectrometry with a fibre from a suspect's clothing.

### 8.8.4  LEAK DETECTION

A leak in a vacuum system can be traced by spraying the outside of the apparatus with helium $(M_r = 4)$. A mass spectrometer tuned to $m/e = 4$ is attached to the apparatus. If any helium atoms are sucked into the apparatus they give rise to a signal.

*... detecting leaks ...*  This technique can detect leaks down to $10^{-12}$ cm$^3$ s$^{-1}$ of gas at 1 atm.

### 8.8.5  SPACE EXPLORATION

The Viking spacecraft carried equipment for gas chromatography combined with mass spectrometry on its voyage to Mars. It analysed the atmosphere of Mars as 95% $CO_2$, 2.5% $N_2$, 1.5% Ar and 1% other gases. The soil of Mars was found to have a

*... space exploration ...*  high content of chlorine and sulphur.

### 8.8.6  POLLUTION MONITORING

The combination of gas chromatography and mass spectrometry can detect pollutants at low concentration. Polychlorinated biphenyls, PCBs, are used as insecticides. They are dangerous substances because they are not biodegradable and because they can

*... pollution monitoring ...*  become part of a food chain leading to high concentrations of the PCBs in animals at the top of the food chain [see DDT, *ALC*, § 20.15]. Low concentrations of PCBs in rivers and lakes can be detected by mass spectrometry. Dioxin, the toxic substance which caused the tragedy of Seveso [see *ALC*, § 30.13], can readily be detected by mass

*... detecting pesticides ...*  spectrometry.

*... analysing oil slicks.*  It is possible to identify oil from a particular oilfield. This can be useful in a case where an oil tanker has illegally washed out its tanks at sea. A sample of the oil slick can be taken for examination. Oil was formed by the decay of marine organisms over millions of years. In addition to hydrocarbons, compounds such as steroids were formed. These compounds are termed **biomarkers**. The content of biomarkers in the oil from a particular oilfield is different from that of all others. The biomarkers in a sample of oil can be separated by gas chromatography and identified by mass spectrometry to establish from which oilfield the sample came and identify the tanker responsible for the pollution.

### 8.8.7  REACTION MECHANISMS

Different isotopes of an element can be detected with ease. This leads to the use of mass spectrometry in isotopic labelling experiments. A well-known example is the elucidation of the mechanism of hydrolysis of carboxylic esters.

*Mass spectrometry is used
in the study of reaction
mechanisms by isotopic
labelling ...*

$$
\underset{\text{OR'}}{\overset{\displaystyle \overset{O}{\parallel}}{R - C}} + H_2O \longrightarrow \underset{\text{OH}}{\overset{\displaystyle \overset{O}{\parallel}}{R - C}} + R'OH
$$

Which bond is broken: the C—O bond or the O—R' bond? If the ester is labelled with $^{18}O$, the question can be answered. The $^{18}O$ turns up in the alcohol, not in the acid; therefore it is the C—O bond which is broken:

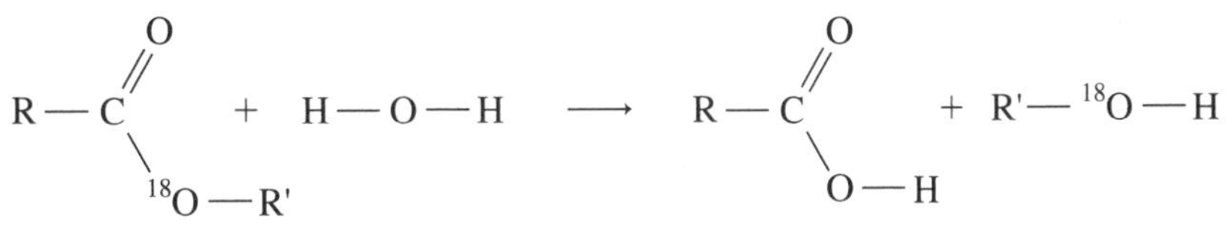

$$R—C \begin{matrix} O \\ \parallel \end{matrix} \quad + \quad H—O—H \quad \longrightarrow \quad R—C \begin{matrix} O \\ \parallel \end{matrix} \quad + \quad R'—^{18}O—H$$

### 8.8.8. GEOLOGICAL DATING

The method of **potassium–argon dating** is used to determine the age of rock samples. The basis of the method is the decay of $^{40}K$ to $^{40}Ar$ (and other products). If a rock contains potassium it will contain a small percentage of $^{40}K$. The $^{40}Ar$ formed by decay may not be able to diffuse out of the crystal structure of the rock. In such a rock, the $^{40}Ar$ content can be used to determine the age of the rock sample. The argon is released by melting the rock in a vacuum and then it is assayed in a mass spectrometer. The method of isotope dilution analysis is used [*ALC*, § 1.9.8] in which a tracer of $^{38}Ar$ is added to the $^{40}Ar$. The potassium content is found by atomic absorption spectroscopy [§ 4.3] or by isotope dilution.

Some $^{40}Ar$ may be present in the rock because air was trapped in it. This can be allowed for. If atmospheric argon is present in the rock, it will contain $^{36}Ar$, $^{38}Ar$ and $^{40}Ar$. The amount of $^{40}Ar$ in the atmosphere is 300 times the amount of $^{36}Ar$. If the amount of $^{36}Ar$ in the rock is measured, the amount of atmospheric $^{40}Ar$ is 300 times this amount. The amount of atmospheric $^{40}Ar$ can now be subtracted from the total amount of $^{40}Ar$ in the rock to give the amount of $^{40}Ar$ due to the decay of $^{40}K$.

*. . . geological dating, e.g. by the potassium–argon method.*

The extent to which $^{40}K$ has decayed to $^{40}Ar$ is then known. From the equation for a first order reaction and the decay constant [*ALC*, § 14.5.3] the age of the rock can be found.

## 8.9 COMBINING TECHNIQUES FOR IDENTIFYING COMPOUNDS

Mass spectrometry gives the relative molecular mass of a compound. It may indicate the number of carbon atoms present and the presence of chlorine and bromine. The fragmentation pattern gives valuable information about the structure of the compound.

Infrared spectrometry may indicate the functional groups present in the sample and will indicate when groups are absent.

*A combination of techniques is often used to identify a compound.*

Visible–ultraviolet spectrometry may give information about chromophores in the molecule.

The NMR spectrum gives information about the structure of hydrogen-containing groups in the molecule.

**Example** Figure 8.9A shows the IR spectrum, the mass spectrum and the NMR spectrum of the compound of molecular formula $C_4H_7NO$. The visible–UV spectrum shows no absorption above 210 nm. What can be deduced about the identity of the compound?

**FIGURE 8.9A**
(a) IR Spectrum of
$C_4H_7NO$, (b) Mass
Spectrum of $C_4H_7NO$,
(c) NMR Spectrum of
$C_4H_7NO$

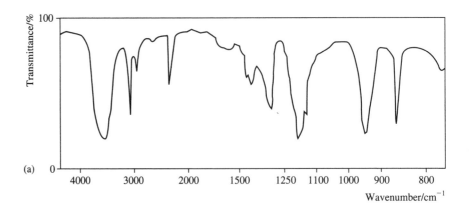

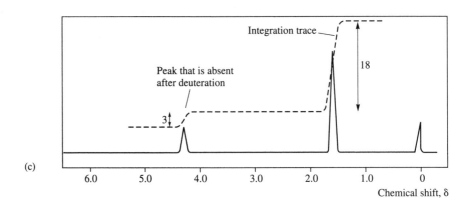

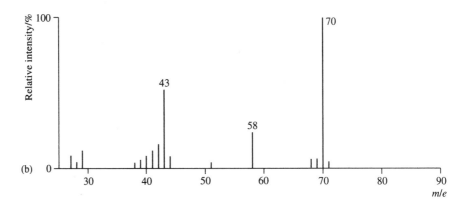

**1.** The IR spectrum shows an absorption peak at 3600–3400 cm$^{-1}$ which is characteristic of the —OH group, a peak at 3000–2900 cm$^{-1}$ which is characteristic of aliphatic C—H bonds and a peak at 2300–2200 cm$^{-1}$ which could be due to a C≡N group.

**2.** The mass spectrum does not show the molecular ion at 85. It has a base peak at 70, which corresponds to the loss of $CH_3$ from the molecule, a peak at 43 which could be due to the stable ion $CH_3CO^+$ and a peak at 58 which could be due to $(CH_3)_2CO^+$.

**3.** The NMR spectrum shows 6 equivalent H atoms and one different H atom. The single H atom has $\delta = 4.3$, which could be an OH group, and it disappears on deuteration, showing that it is a labile H atom. With $\delta = 1.5$, the 6 H could be 2 $CH_3$ groups.

Putting all the information together, the compound is 2-hydroxy-2-methylethanenitrile (the cyanhydrin of propanone).

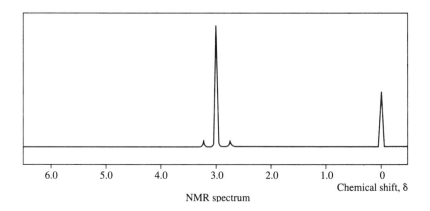

==== CHECKPOINT 8.9 ====

**1.** The compound of molecular formula $C_4H_4O_3$ has the spectra shown below. The visible–UV spectrum shows no absorption above 210 nm. What can you deduce about the identity of the compound?

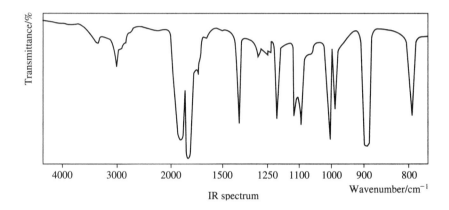

IR spectrum

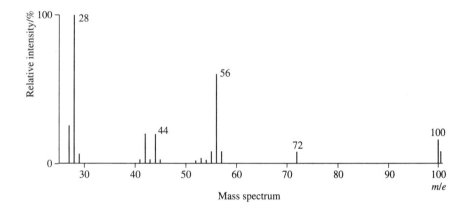

Mass spectrum

NMR spectrum

**QUESTIONS ON CHAPTER 8**

**1.** There are four esters of formula $C_4H_8O_2$.

(*a*) Write the formulae of the four compounds.

(*b*) Which of the four gives the mass spectrum below?

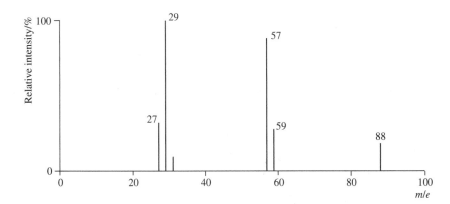

**2.** The figure shows a simplified version of the mass spectrum of ethanol, $C_2H_5OH$. Explain the origin of the six peaks.

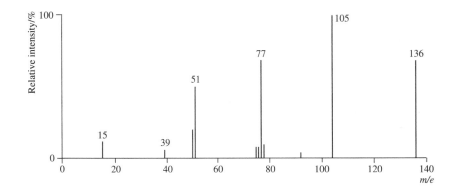

**3.** Identify the compound $C_8H_8O_2$ from its mass spectrum.

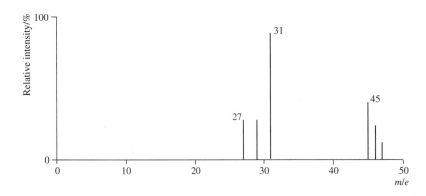

**4.** Identify the compound $C_3H_6O$ from its mass spectrum.

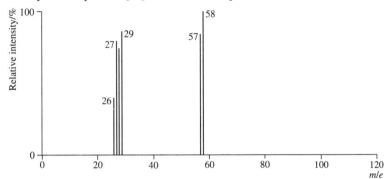

**5.** The mass spectrum of a compound $C_6H_{14}O$ is shown. From the smell and the high volatility the compound appears to be an ether. Identify the peaks in the spectrum and identify the compound as far as you can.

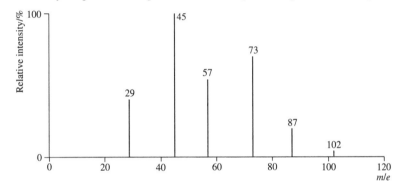

## QUESTIONS ON COMBINED TECHNIQUES

**1.** (*a*) Suggest the identity of the ions responsible for the major peaks in the mass spectrum below. Try to identify the compound.

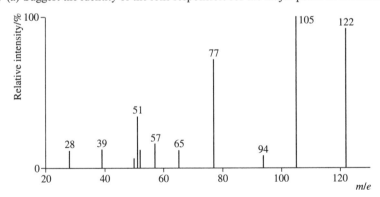

(*b*) How does the infrared spectrum shown below support your answer?

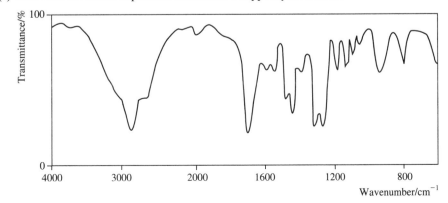

**2.** Identify the compound which gave the following mass spectrum and infrared spectrum.

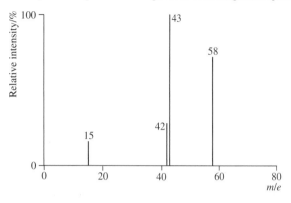

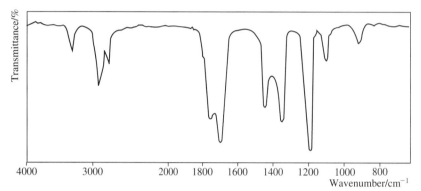

**3.** The IR spectrum and the mass spectrum of a compound are shown.

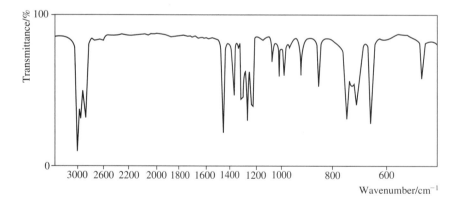

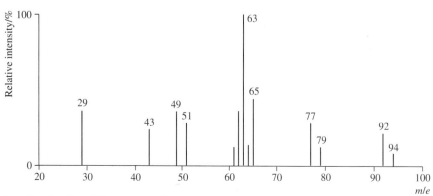

Deduce the identity of the compound.

**4.** The infrared and mass spectra of a compound are shown in the figures. Deduce as much as you can about the identity of the compound.

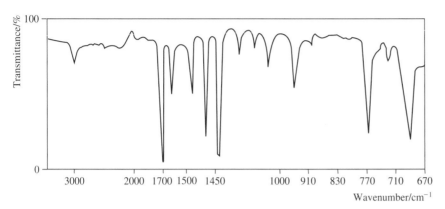

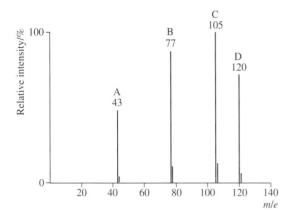

**5.** (*a*) (i) Define the terms *relative molecular mass* and *isotope*.
(ii) Calculate the relative atomic mass of carbon which contains two main isotopes $^{12}C$ (98.9%) and $^{13}C$ (1.10%). Give your answer to an appropriate number of significant figures.

(*b*) The diagram below shows a simplified mass spectrum for an organic compound **X** which contains carbon, hydrogen and oxygen only. The infrared spectrum of **X** indicates the presence of a C=O group.
(i) Deduce from the mass spectrum the relative molecular mass of **X**.
(ii) Account for the very small peaks that occur next to each of the peaks A, B, C and D at masses of 44, 78, 106 and 121 respectively.

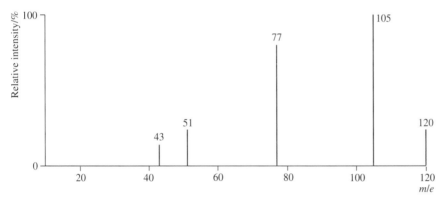

(*c*) (i) Estimate the number of carbon atoms present in ion D and hence explain why compound **X** is likely to be aromatic and to have the formula $C_6H_5COCH_3$.
(ii) Deduce the formulae of the ions responsible for peaks B and C.
(iii) Suggest a reason for the absence of a recorded peak at $m/e = 15$.

(*d*) Give the reagents and the result of a simple chemical test to show the presence of the carbonyl group in X.

[L]

**6.** The following spectra were obtained for compound **R**, $C_7H_8O$

Use the spectra to identify the functional groups present in the compound and to deduce its structure. You should identify all relevant absorptions by which you arrive at your answer.

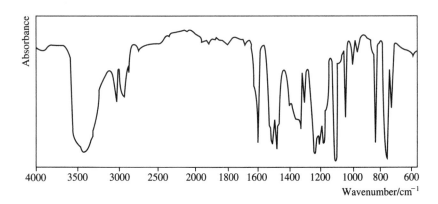

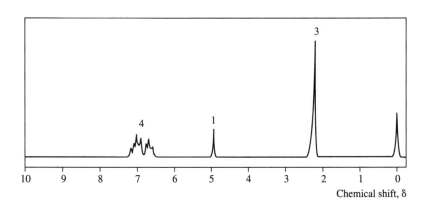

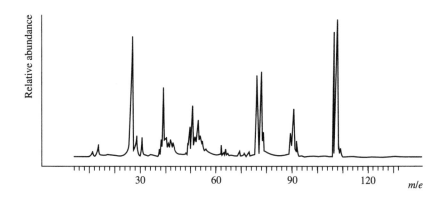

[C]

**7.** (*a*) (i) Draw and label a diagram of a mass spectrometer.
(ii) Explain how a mass spectrometer works.
The figure is a simplified diagram of a mass spectrum of ethanol, $C_2H_5OH$.

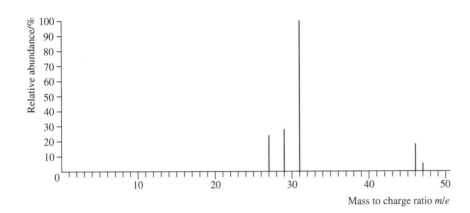

(iii) Identify the molecular ion peak and the $M + 1$ peak by their $m/e$ values and explain the origin of the $M + 1$ peak.
(iv) Suggest identities for the fragments that could correspond to the remaining peaks on the spectrum.

(*b*) The figure below is a diagram of a low-resolution proton nuclear magnetic resonance spectrum of ethanol.

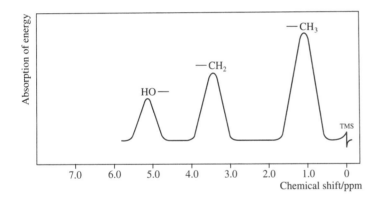

With reference to this spectrum explain:
(i) the origin of the peaks
(ii) the significance of the area under the peaks
(iii) what is meant by chemical shift

[C]

**8.** (*a*) Consider the following instrumental techniques used in structural analysis.
infrared spectroscopy
ultraviolet/visible spectroscopy
proton magnetic resonance spectroscopy
mass spectrometry
X-ray diffraction
(i) Which of these techniques uses electromagnetic radiation of the lowest frequency?
(ii) Which technique uses electromagnetic radiation of the shortest wavelength?
(iii) In which technique is electromagnetic radiation absorbed by vibrating bonds?
(iv) Which technique is the most generally appropriate for examining isotopically labelled molecules?
(v) Which technique can be used to determine the concentration of a compound in solution down to levels of approximately $10^{-5}$ mol $l^{-1}$?
(vi) Indicate which technique often makes use of deuterated solvents and explain why this is so.

(*b*) The visible absorption spectrum of the water effluent from a textile dyeworks is shown below. The effluent contains two dyes **X** and **Y**, whose absorption bands are shown.

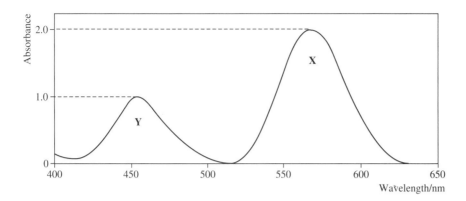

(i) Explain the process whereby a molecule absorbs energy from light.

(ii) Define the *absorbance* ($A$) of a solution in terms of the incident light intensity ($I_0$) and the emergent light intensity ($I$).

(iii) State the expression which relates the molar extinction (absorption) coefficient ($\varepsilon$) of a substance to the absorbance of its solution at $\lambda_{max}$, the molar concentration ($c$), and the pathlength ($l$) of the solution.

(iv) Deduce from the above spectrum the *relative* concentrations of X and Y in the effluent, given that the $\varepsilon_{max}$ value for X is three times that of Y.

(v) On a particular occasion it was found that, in the absence of X and Y, the effluent was brown in colour. State why the absorbance measurements would give erroneous concentration values for X and Y when these two compounds are present in such an effluent. How might the error be minimised if a sample of the effluent containing no X or Y were available?

[N]

**9.** (*a*) (i) Draw a simple diagram of a mass spectrometer, labelling clearly the major components of the system. Draw the paths through the apparatus of a heavy ion and a light ion at a given intensity of the magnetic field.

(ii) In which region of the instrument does ion fragmentation take place?

(iii) What can be deduced about the composition of a compound if it shows two molecular ion peaks of equal intensity?

(*b*) A compound is known to have either structure **A** or **B**, and is investigated by infrared spectroscopy and proton magnetic resonance spectroscopy.

$$CH_3CH_2\!-\!\bigcirc\!-\!OH \qquad\qquad CH_3\!-\!\bigcirc\!-\!O\!-\!CH_3$$

**A**                                                             **B**

(i) The infrared spectrum of the compound measured in solution in trichloromethane shows a peak at a wavenumber of $3600\ cm^{-1}$, the appearance of which is strongly dependent on the concentration of the solution. With this information, assign structure **A** or **B** to the compound and give an appropriate explanation.

(ii) The proton magnetic resonance spectrum shows (with other peaks) a triplet peak and a quartet peak system. State which groups of protons in the compound are responsible for these two multiplets.

(iii) Explain how deuterium oxide could be used to distinguish between structures **A** and **B** by proton magnetic resonance spectroscopy.

[N]

**10.** In solution, pentane-2,4-dione exists in two isomeric forms, **A** and **B**. The proportions of **A** and **B** in the equilibrium mixture depend on temperature and solvent.

$$\begin{array}{ccc}
O\!=\!C\diagdown^{CH_3} & & O\!-\!C\diagdown^{CH_3} \\
\ \ \ \diagdown_{CH_2} \rightleftharpoons & H & \ \ \ \diagdown_{CH_2} \\
O\!=\!C\diagup & & O\!=\!C\diagup \\
\ \ \ \diagdown_{CH_3} & & \ \ \ \diagdown_{CH_3}
\end{array}$$

**A**                                                             **B**

(a) (i) Predict which of the two isomers **A** or **B** gives rise to the absorption band of longest wavelength in the ultraviolet region of the spectrum. State the structural feature of the molecule responsible for absorption at this wavelength.
(ii) Assuming that the pentane-2,4-dione concentration, absorbance and cell path length are known for the spectroscopic measurement in (i) above, what additional quantity would be required in order to calculate the value of the equilibrium constant for the conversion of **A** into **B**?

(b) The infrared absorption spectrum of an equilibrium mixture of **A** and **B** shows a band due to the hydroxyl group at a lower frequency than normally found for the hydroxyl group. Suggest an explanation for this observation.

(c) The proton magnetic resonance spectrum of pentane-2,4-dione measured for the pure liquid is shown below. Analysis of the spectrum shows that both forms, **A** and **B** are present in equilibrium.

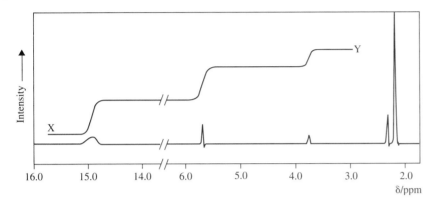

(i) Addition of deuterium oxide, $D_2O$, to the sample results in the disappearance of the peak at $\delta = 14.9$. Indicate which proton or group of protons in either **A** or **B** is responsible for the peak at $\delta = 14.9$, and explain the effect of $D_2O$.
(ii) Explain the significance of trace X–Y and use it to determine the relative concentrations of **A** and **B** in the mixture, given that the peak at $\delta = 3.75$ corresponds to the central $CH_2$ protons of form **A**.

(d) Many azo dyes also show equilibria like **A** ⇌ **B** in solution, but in the solid state they generally exist exclusively in one form or another. If a crystal of such an azo dye were available, indicate a technique that would enable the actual isomeric structure of the molecules in the crystal to be determined. What *two* measurements are required to obtain the complete structure of the compound?

[N]

**11.** A sample of mineral water from a plastic bottle was tested for contamination by compound **C**.

(a) In order to identify **C** by infrared spectroscopy, it was first extracted from the water and concentrated to give an oily liquid.
(i) Give *one* reason why the substance was extracted from the water before the infrared spectrum was measured.
(ii) If a sample of pure **C** were available, how would infrared spectroscopy be used to confirm the structure of the extracted material?
(iii) The extracted oil contains residual water, which shows three well-defined absorption peaks in the infrared spectrum. Draw simple diagrams to show the types of vibration which give rise to the three absorption peaks.
(iv) What is the characteristic of a vibrational mode in a water molecule which ensures that infrared radiation can be absorbed?

(b) The mass spectrum of the extract in (a) shows a molecular ion with $m/e = 222$.
(i) Explain briefly how ions are produced in the mass spectrometer.
(ii) Explain briefly how the resulting ions are separated and analysed in the mass spectrometer.
(iii) The molecular ion of $m/e = 222$ corresponds not only to the molecular formula of **C** but also to other formulae (e.g. $C_{13}H_{18}O_3$). Explain how measurement of the value of $m/e$ may be used to distinguish between such molecular formulae.

(c) The proton magnetic resonance spectrum of **C** is measured in solution in deuterotrichloromethane, using tetramethylsilane as an internal reference.
(i) Why does tetramethylsilane show a single resonance peak?
(ii) The ethyl group protons in **C** appear as characteristic quartet and triplet signals in the spectrum. Explain with the aid of a simple diagram why the $CH_2$ protons in the ethyl group appear as a quartet.

[N]

# APPENDIX: VOLUMETRIC ANALYSIS

**Titrimetric methods of analysis** measure the amount of a reagent that is consumed by the analyte (the substance being analysed).

**Volumetric analysis** measures the volume of a solution of known concentration that is needed to react completely with the analyte. Volumetric analysis is widely used because the methods employed are accurate, convenient and readily automated. The techniques of volumetric analysis have been described in *A-Level Chemistry*, and can be found under the following references.

## STANDARD SOLUTIONS

Concentration of solution, § 3.13.

Preparing a standard solution by weighing, § 3.13.1, and by dilution, § 3.13.2.

## TITRATION

Method of carrying out a titration, § 3.14.

Back-titration, § 3.14.1.

## ACIDS AND BASES

Theory of acids and bases, § 12.7.1.

Theory of acid–base titrations, § 12.7.9.

## REDOX REACTIONS

Theory of redox reactions, § 3.15.1.

Balancing equations for redox reactions, § 3.16.3.

Redox reactions in titrimetric analysis, § 3.18.

## CONDUCTIMETRIC TITRATION, § 12.7.10.

The conductance of a solution changes during the course of titration. This method is used for acid–base titrations.

## POTENTIOMETRIC TITRATION, § 13.2.2.

Potentiometric titration is used to follow neutralisation reactions. The potential of a glass electrode depends on the pH of the solution surrounding it and changes during the course of neutralisation. The change is very rapid on each side of the equivalence point.

Potentiometric titration can be used for redox titrations. A platinum electrode is immersed in a solution of one of the reagents and coupled with a reference electrode. The potential of the platinum electrode, and therefore the e.m.f. of the cell, changes as titration proceeds, and changes rapidly when the equivalence point is reached.

### PRECIPITATION TITRATION, § 12.7.16.

Precipitation titration can be used for e.g. the estimation of halides by titration against a standard solution of silver nitrate. A precipitate of silver chloride or bromide or iodide forms. The equivalence point is detected by means of an indicator, e.g. potassium chromate(VI), which gives a coloured precipitate with silver ions after all the halide ions have been precipitated.

### COMPLEXOMETRIC TITRATION, § 12.7.17.

A number of metal ions form very stable complexes with edta. Ions such as $Ca^{2+}$, $Mg^{2+}$ and $Zn^{2+}$ can be estimated by titration against a standard solution of the disodium salt of edta. The equivalence point is detected by adding an indicator, e.g. Eriochrome Black T, which forms a coloured complex with the metal ion. At the equivalence point the colour of the free indicator is seen.

### CALCULATIONS

Further practice with calculations on volumetric analysis can be found in *Calculations for A-Level Chemistry* by E N Ramsden (Stanley Thornes).

# ANSWERS

The answers to the examination questions are the responsibility of the author. The examination boards take no responsibility for them.

## CHAPTER 1: CHROMATOGRAPHY

**Checkpoint 1.4**
1. (a) See § 1.1
   (b) See § 1.1
   (c) See § 1.2
   (d) See § 1.1
   (e) See § 1.3
   (f) See § 1.3
   (g) See § 1.1
2. (a), (b) See § 1.3
3. (a) Either there is only one substance in the sample or the components travel at the same rate in the solvent chosen.
   (b) Try a different solvent.

**Checkpoint 1.6**
1. (a) A carboxylic acid with a large alkyl group, e.g. $C_{16}H_{31}CO_2H$.
   (b) Pass the mixture through a GC column. Identify the peaks; see Figure 1.5E. Pass the eluate to a mass spectrometer. Find the molar mass of each peak. All have the formula $C_nH_mCO_2CH_3$, where the alkyl group can be saturated or unsaturated, therefore from $M_r$ the formula can be found.
2. (a), (b), (c) See §§ 1.5.2 and 1.5.3
   (d) See § 1.5.1
3. (a) The particle size is smaller, therefore separation takes place over a shorter distance.
   (b) High pressure has to be used to drive the eluant through the densely packed column, therefore the column must be strongly constructed, e.g. of stainless steel.
   (c) Compounds which decompose at the higher temperatures needed for GC and inorganic compounds
   (d) Rapid, good separation, less expensive, easier to interface with mass spectrometers and other detectors

**Checkpoint 1.8**
1. $K^+$ and $Sr^{2+}$
2. (a) See § 1.7.1
   (b) Cation exchange: e.g.
   $$2RSO_3^- H^+(s) + Ca^{2+}(aq) \rightleftharpoons (RSO_3^-)_2Ca^{2+}(s) + 2H^+(aq)$$
   Anion exchange: e.g.
   $$RN(CH_3)_3^+ OH^-(s) + Br^-(aq) \rightleftharpoons RN(CH_3)_3^+ Br^-(s) + OH^-(aq)$$
3. See § 1.7.1
4. See § 1.7.3
5. See § 1.7.4

6. Peak A: These molecules were too large to enter the pores and were washed through with a smaller volume of eluant.
7. Amino acid molecules are much smaller than protein molecules and pass unhindered through the column.

**Questions on Chapter 1**
1. See § 1.3
2. (a) For $R_F$ see § 1.3
   (b) Try a different eluant.
3. Dry the chromatogram and run a second chromatogram at right angles in a different solvent.
4. (a) Inexpensive, fairly rapid, efficient – gives a good separation in a few cm, can be run in two dimensions
   (b) (i) See § 1.5.1
   (ii) Scrape off the area of the layer that contains a component, dissolve the component, filter to remove the adsorbent, evaporate the solution to obtain the component.
5. (a) See § 1.5
   (b) (i) TCD do not consume the sample, but the sensitivity is not high.
   (ii) FID are very sensitive but they consume the sample.
   (iii) ECD are very sensitive and do not consume the sample, but they can only be used on compounds which capture electrons, e.g. halogenocompounds.
6. Separate the steroids in a specimen of urine from other compounds. Run a GC on the steroids. Compare the retention times of the steroids with those of forbidden steroids measured under the same conditions.
7. (a) efficient separation
   (b) small particle size
8. (a) e.g. $2R^- Na^+(resin) + Ca^{2+}(aq) \rightleftharpoons$
   $$(R^-)_2Ca^{2+}(resin) + 2Na^+(aq)$$
   $RN(CH_3)_3^+ OH^- (resin) + I^-(aq) \rightleftharpoons$
   $$RN(CH_3)_3^+I^-(resin) + OH^-(aq)$$
   $2RSO_3^- H^+(resin) + Ca^{2+}(aq) \rightleftharpoons$
   $$(RSO_3^-)_2Ca^{2+}(resin) + 2H^+(aq)$$
   (b) See § 1.7.2 (c) See §1.7.4
9. 9 min = chloroethanoate, 17 min = nitrate, 25 min = citrate
10. (a) There are at least four, but a peak could represent two or more proteins with molecules of the same size.
    (b) A < B < C < D
    (c) Try to separate the component by gel filtration on a column with smaller gel particles which give better separation. Try a different method, e.g. electrophoresis; see § 2.1.

## CHAPTER 2: ELECTROPHORESIS

**Checkpoint 2.2**
1. Band 5 – have moved furthest
2. Band B – have moved further

**Checkpoint 2.3**
1. Jon
2. F. Green

**Questions on Chapter 2**
1. See § 2.1
2. See § 2.3.3
3. (a) See § 2.2
   (b) DNA, § 2.2 and proteins, § 2.3.1
4. See § 2.4

## CHAPTER 3: LIGHT

**Checkpoint 3.2**
1. (a) yellow-green  (b) blue-green  (c) orange
   (d) blue-violet  (e) violet  (f) red
2. (a) 620–650 nm  (b) 510–590 nm
3. (a) 430–650 nm  (b) 395-430 nm and 500–650 nm

**Checkpoint 3.5**
1. A is yellow because it absorbs at 400–480 nm (violet to blue-green), B is red because it absorbs at 450–550 nm (blue to green), C is blue because it absorbs at 620–680 nm (yellow to red).
2. (a) red  (b) violet
3. (a) radio waves  (b) cosmic rays
4. (a) (i) 670–3300 cm$^{-1}$  (ii) $2.0 \times 10^{13}$–$1.0 \times 10^{14}$ Hz
   (b) (i) $3.3 \times 10^3$–$5.0 \times 10^4$ cm$^{-1}$ or 3300–50 000 cm$^{-1}$
   (ii) $1.0 \times 10^{14}$–$1.5 \times 10^{15}$ Hz

**Questions on Chapter 3**
1. See § 3.1
2. (a) orange  (b) violet  (c) red
3. (a) See § 3.3  (b) The absorption spectrum arises from transitions between electron levels. These are different for every type of atom.
4. (a) See § 3.4  (b) Supply energy by e.g. a flame or an electric discharge.
5. (a) See § 3.6  (b) see § 3.6
6. Divide $3.0 \times 10^8$ m s$^{-1}$ by $\lambda$/m.
   (a) $3.63 \times 10^{13}$ Hz  (b) $1.2 \times 10^8$ Hz
   (c) $4.32 \times 10^{14}$ Hz
7. Use $\lambda = c/v$. With $c$ in m s$^{-1}$ and $v$ in s$^{-1}$, $\lambda$ will be in m.
   (a) 2.50 m  (b) 2.86 m  (c) $2.0 \times 10^{-4}$ m = 0.20 mm
   (d) $1.0 \times 10^{-9}$ m = 1.0 nm
8. (a) left to right  (b) left to right  (c) see § 3.6
   (d) All involve transitions down to the same energy level.
   (e) The orbitals become closer together at higher energy; see Figure 3.6E.
   (f) The electron has received so much energy that it has left the orbitals; the atom has ionised.
9. At high temperature, the atom has more energy and the electron can occupy higher orbitals. Transitions take place between a larger number of orbitals.

## CHAPTER 4: ATOMIC SPECTROSCOPY

**Checkpoint 4.6**
1. See §§ 4.2, 4.3. The energy of the electron transition corresponds to the frequency of yellow light.
2. The electron transition takes place in the atom or ion, independently of any anions present.
3. A source of radiation; see § 4.6.

**Questions on Chapter 4**
1. The dark lines in the AAS are at the same wavelengths as the bright lines in the AES. Absorption occurs when an electron is promoted to a higher orbital. Emission occurs when the electron returns to its ground state. The transitions are equivalent to the same energy difference in both cases, therefore the wavelength is the same in both cases.
2. See § 4.6
3. (a) A source of radiation
   (b) A hollow cathode lamp is a monochromatic source of radiation; see Figure 4.6B.
4. See § 4.7
5. (a) See § 4.8
   (b) Let [Pb] = $x \, \mu$g dm$^{-3}$
   $x \propto 0.075$, and $(6.0 + x) \propto 0.115$
   $x/(6.0 + x) = 0.075/0.115$
   $x = 11.25 \, \mu$g dm$^{-3}$
6. (b) Average of samples A, B and C = 35.8. On the graph, this corresponds to 31 $\mu$g Na cm$^{-3}$ = $3.1 \times 10^{-3}$%.
7. 0.500 ppm Pb
8. See § 4.9

## CHAPTER 5: VISIBLE–ULTRAVIOLET SPECTROMETRY

### Answers to Checkpoint 5.4

1. C: $A = \log(100/25) = 0.602$
   D: $A = \log(100/75) = 0.124$
2. (a) From $A = \log(100\%/\%$ transmitted$)$,
   for E, $A = 75\%$ and for F, $A = 18.8\%$
   (b) 56.2% and 3.5%
3. $A = \varepsilon cl$ therefore $\varepsilon = 0.691/(1.00 \times 2.84 \times 10^{-5})$
   $= 2.43 \times 10^4$ at 520 nm

### Checkpoint 5.5

1. A absorbs violet (380–430 nm) and appears yellow.
   B absorbs red (480–520 nm) and is green.
   C absorbs orange (600–650 nm) and is blue-violet.
2. A plot of absorbance against volume of $M^{2+}$ (aq) or
   against volume of L (aq) gives a maximum at volume of
   $M^{2+}$ (aq) $= 3.3 \text{ cm}^3$, volume of L (aq) $= 6.6 \text{ cm}^3$. Ratio
   $[M^{2+} \text{(aq)}] / [L \text{(aq)}] = 1/2$, and formula is $ML_2$.

### Checkpoint 5.9

1. $\sigma$ and $\pi$
2. Benzene absorbs in the UV. The —$NO_2$ group is an
   auxochrome; it extends the delocalised system of double
   bonds in benzene.
3. The azo group links two benzene rings by a system of
   delocalised bonds. This increases the ease of promotion
   and the absorption of radiation.
4. $a, b, c, d$, yes; $e$, no; $f$, yes
5. —$OCH_3$, non-bonding orbitals on O and $\sigma$ orbitals also
   —$NH_2$, non-bonding orbitals on N and $\sigma$ orbitals also
   —$CN$, $\pi$ orbitals in —$C{\equiv}N$ and $\sigma$ orbitals
   —$CHO$, $\pi$ orbitals in $>C{=}O$ and $\sigma$ orbitals
   —$NO_2$, $\pi$ orbitals in —$N{=}O$ and $\sigma$ orbitals
   $\searrow O$

### Checkpoint 5.10

1. Evaporate to dryness, drive off water of crystallisation by
   heating to 100 °C. The anhydrous $CuSO_4$ that remains is
   colourless.
2. Thiocyanate absorbs blue-green light, 500 nm; Prussian
   blue absorbs orange, 610 nm.
3. (a) green  (b) purple
   (c) It promotes an electron from a low-energy d orbital to
   a high-energy d orbital.
4. Zn has the electron configuration $(Ar)3d^{10}4s^2$; therefore
   $Zn^{2+}$, (Ar) $3d^{10}$, has a full d subshell of electrons. Cu has
   the electron configuration $(Ar)3d^{10}4s^1$; therefore $Cu^+$,
   $(Ar)3d^{10}$, has a full d subshell of electrons.

### Checkpoint 5.11

1. (a) absorbs (A) 200–300 nm  (B) 200–300 and 350–400
   and 500–590 nm
   (b) (A) no visible light absorbed  (B) green and yellow
   visible light absorbed
   (c) (A) appears colourless  (B) red-violet light
   transmitted; looks purple
   (d) The system of conjugated double bonds in
   phenolphthalein in alkaline solution permits extensive
   delocalisation of electrons and is therefore a
   chromophore.
2. Concentration of $NO_3^-$ in water $= x$ ppm
   (Absorbance of water/Absorbance of sample) $=$
   $54.8/64.5 = x/1.00$
   Therefore $x = 0.850$ ppm
3. Absorbance 0.700 results from 5.00 μg; absorbance 0.840
   results from 6.00 μg
   Concentration $= 6.00 \text{ μg}/10 \text{ dm}^3 = 600 \text{ μg m}^{-3}$.

4. Absorbance 0.316: Concentration $= 1.00 \text{ mg dm}^{-3}$
   Absorbance 0.296: Concentration $= 0.747 \text{ mg dm}^{-3}$
   Therefore wastewater has $(250/10.0) \times 0.747$
   $= 18.7 \text{ mg dm}^{-3}$ of Mn

### Checkpoint 5.12

1. (a) less energy
   (b) conversion into kinetic energy; see § 5.12
2. phosphorescence; see § 5.12
3. See § 5.12

### Questions on Chapter 5

1. (a) $\pi \longrightarrow \pi^*$
   (b) $n \longrightarrow \sigma^*$ and $n \longrightarrow \pi^*$
   (c) $\pi \longrightarrow \pi^*$
   See § 5.8
2. (a) $[Cu(H_2O)_6]^{2+}$, $[Cu(NH_3)_4(H_2O)_2]^{2+}$
   $[Cu(H_2NCH_2CH_2NH_2)_2(H_2O)_2]^{2+}$
   (b) $\Delta E$ for en $> \Delta E$ for $NH_3 > \Delta E$ for $H_2O$
3. (a) chromophore
   (b) See § 5.7
   (c) $\pi$-electrons or n-electrons; see § 5.8
   (d) electronic transitions from bonding and non-bonding
   orbitals to antibonding orbitals; $\pi \longrightarrow \pi^*$ and
   $n \longrightarrow \sigma^*$ and $n \longrightarrow \pi^*$
   (e) See § 5.9
4. (a) See § 5.10 for ligand field theory.
   (b) These ions have a full 3d shell.
5. See § 5.9: —$NO_2$ is an auxochrome, forming a delocalised
   electron system with the benzene ring.
6. (a) $A = \varepsilon \times 6.24 \times 10^{-5} \times 1.00 = 0.437$
   $\therefore \varepsilon = 7.00 \times 10^3$
   (b) $A = \log(I_0/I)$  $\therefore I = 36.5\%$
   (c) $A = 0.0625 = \varepsilon cl = 7.00 \times 10^3 \times 2.50$
   $\therefore c = 3.57 \times 10^{-6} \text{ mol dm}^{-3}$
   In the original solution, $c = 7.14 \times 10^{-5} \text{ mol dm}^{-3}$.
7. (a) (i) $\pi \longrightarrow \pi^*$ and $n \longrightarrow \pi^*$  (ii) $\pi \longrightarrow \pi^*$
   (b) A conjugated system of double bonds increases
   the extent of delocalisation and decreases the
   energy needed for promotion of an electron. Thus,
   since $\Delta E = h\nu$, the frequency of light absorbed
   decreases and the absorption shifts to longer
   wavelengths.
   (c) (i) See discussion of ligand field theory, § 5.10.
   (ii) The electron is promoted to a d orbital of higher
   energy. The value of $\Delta E$ corresponds to light of
   wavelength in the region 550–600 nm and the
   absorption of this light leaves the complementary
   colour violet.
8. (a) The ions of transition metals have incomplete d shells.
   In a complex ion, ligand field splitting makes it
   possible for electron transitions to take place from
   one d orbital to another d orbital of higher energy;
   see § 5.10.
   (b) (i) A, red; B, blue-violet
   (ii) The longer wavelength in B means that $\nu$ is lower.
   Since $\Delta E = h\nu$, $\Delta E$ in B is less than in A.
   (iii) In solution C, the concentrations of L and J are
   equal. Solutions A and B show that the molar
   absorbance of $CrJ_6$ is greater than that of $CrL_6$. In
   solution C, the absorbance of $CrL_6$ is greater;
   therefore $CrL_6$ predominates over $CrJ_6$: L forms the
   stronger bond with $Cr^{3+}$.

# CHAPTER 6: INFRARED SPECTROMETRY

## Checkpoint 6.6

1. There is a C—H aliphatic absorption at $3000\,cm^{-1}$ and a C—Cl absorption at $765\,cm^{-1}$. The compound is a chloroalkane. (In fact this is the IR spectrum of $CHCl_3$.)

2. C≡N at $2250\,cm^{-1}$ and C—H aliphatic at $3000\,cm^{-1}$ and C—H vibrations at 1460 and $1430\,cm^{-1}$. The presence of C≡N and the absence of aromatic C—H and aromatic C—C indicate an aliphatic nitrile. (In fact this is propanenitrile, $CH_3CH_2CN$.)

3. C—H aromatic above $3000\,cm^{-1}$, C—H of CHO at 2700 and $2800\,cm^{-1}$, C=O at $1700\,cm^{-1}$, C—C aromatic at $1600\,cm^{-1}$ and C—H at $1200\,cm^{-1}$. The spectrum is that of an aromatic aldehyde. (In fact it is that of benzaldehyde.)

4. C—H aliphatic at $3000\,cm^{-1}$, C—H of CHO at 2720 and $2830\,cm^{-1}$, C=O at $1730\,cm^{-1}$. The compound is an aliphatic aldehyde. (In fact it is propanal.)

5. C≡N at $2200\,cm^{-1}$, C—H aromatic at $3050\,cm^{-1}$, C—C aromatic at $1480\,cm^{-1}$. There are indications of an aromatic ring and a nitrile group. The simplest compound possessing these structures is benzonitrile or benzenecarbonitrile, $C_6H_5CN$. One could identify the compound by a molar mass determination. (In fact the spectrum is that of benzenecarbonitrile.)

6. C=O at $1700\,cm^{-1}$ and O—H with hydrogen bonding at $2600-3400\,cm^{-1}$ suggest a carboxylic acid. Since it is a liquid, it must be an aliphatic acid. (In fact the spectrum is that of ethanoic acid.)

## Questions on Chapter 6

1. The O—H absorption is present at $3450\,cm^{-1}$, and the C—O absorption at $1050\,cm^{-1}$, and C—H aliphatic absorption at $2900\,cm^{-1}$, suggesting an aliphatic chain. There is an absence of bands below $1000\,cm^{-1}$ which would indicate an aromatic compound. It looks like an aliphatic hydrocarbon chain with the bonds C—O and O—H, that is an alcohol. (In fact this is the spectrum of dodecanol, $CH_3(CH_2)_{10}CH_2OH$.)

2. The C=O absorption at $1715\,cm^{-1}$ and $CH_2$ at $1460\,cm^{-1}$ can be seen. The absence of a big absorption band below $1000\,cm^{-1}$ indicates an aliphatic compound. The indications are an aliphatic compound with a carbonyl group. The C=O group is at the lower end of the absorption range and is probably a ketone. From $M_r = 98$, subtract 28 for CO, leaving 70. This could be $C_5H_{10}$ and the compound could be $C_6H_{10}O$, cyclohexanone.

3. The C—Cl absorption is present at $750\,cm^{-1}$. There are aliphatic C—H vibrations at $2940\,cm^{-1}$. There is no evidence of an aromatic ring at $800-700\,cm^{-1}$. This seems to be a chloroalkane. (It is in fact dichloromethane.)

4. The C=O absorption is seen at $1700\,cm^{-1}$, C=C at $1400\,cm^{-1}$, C—H at $1100-70\,cm^{-1}$ and at $3000\,cm^{-1}$, O—H at $2500-2800\,cm^{-1}$. An aromatic compound with a C=O group and an O—H group, e.g. an aromatic carboxylic acid. (In fact this is benzoic acid.)

5. The C=O absorption is seen at $1725\,cm^{-1}$, C—H aromatic at $3000\,cm^{-1}$ and C—O—C aliphatic at $1100\,cm^{-1}$. An aromatic ester would fit the spectrum. (In fact this is ethyl benzoate.)

6. The peak at $3100-3600\,cm^{-1}$ is due to $NH_2$. The C—H absorption at $2800-2900\,cm^{-1}$ is due to an aliphatic compound: for aromatic compounds the C—H wavenumber is $3000-3150\,cm^{-1}$. The spectrum is that of an aliphatic amine. (In fact it is that of 1-amino-2,2-dimethylpropane.)

7. (a) HCl has a dipole moment; see §6.1.
   (b) Molecular vibrations increase in amplitude; see §6.1.
   (c) (i) $2200\,cm^{-1}$, C≡N; $3000\,cm^{-1}$, C—H aliphatic; $1680-1640\,cm^{-1}$, C=C aliphatic
       (ii) $C_3H_3N$, $CH_2$=CH—C≡N

8. (a) There are many bending and stretching vibrations; see §6.1.
   (b) Different plastics contain different bonds. In addition to C—C and C—H, polyesters contain C—O and C=O, polyamides contain N—H and C=O.
   (c) (i) For **A** use the absorption at $3100\,cm^{-1}$ due to the aromatic C—H bond; for **B** use the absorption at $3400\,cm^{-1}$ due to N—H; for **C** use the absorption at $1725-1750\,cm^{-1}$ due to C=O.
       (ii) **A** could be polystyrene (with the aliphatic C—H at $3000-2850\,cm^{-1}$ and the aromatic C—H at $3150-3000\,cm^{-1}$. **B** could be a polyamide (with C=O at $1750-1650\,cm^{-1}$ and N—H at $3300-3400\,cm^{-1}$). **C** could be a polyester (with C=O at $1750\,cm^{-1}$ and C—O at $1150-1200\,cm^{-1}$).

# CHAPTER 7: NUCLEAR MAGNETIC RESONANCE SPECTROMETRY

## Checkpoint 7.7

1. Ethanol, $C_2H_5OH$. The —$CH_3$ 1:2:1 triplet at $1.0-1.3\,\delta$ and the —$CH_2$— 1:3:3:1 quartet at $3.5-3.9\,\delta$ are recognisable. The peak at $5.2\,\delta$ is due to —OH and is absent in $^2H_2O$ because the —OH proton is labile.

2. 1-amino-2,2-dimethylpropane, $(CH_3)_3CCH_2NH_2$. The —$CH_3$ absorption at $0.9\,\delta$ is strong. The peak at $2.5\,\delta$ could be due to R—$CH_2$—R and that at $1.0\,\delta$ to $RNH_2$. The ratio 9 H in $CH_3$ groups : 2 H in —$NH_2$ groups : 2 H in —$CH_2$— groups makes the compound $(CH_3)_3CCH_2NH_2$ a possibility.

3. Propanoic acid, $CH_3CH_2CO_2H$. The —$CH_3$ and —$CH_2$ groups are recognisable – as in Question 1. The peak at $11.7\,\delta$ could be due to —$CO_2H$, which does not give a peak in $^2H_2O$ because the H is labile. The ratio 3 H in —$CH_3$ : 2 H in —$CH_2$— : 1 H in —$CO_2H$ identifies the compound as $CH_3CH_2CO_2H$.

4. Ethylbenzene, $C_6H_5C_2H_5$. The $CH_3CH_2$— group is recognisable as in Question 1. The peak at $7.2\,\delta$ is due to $C_6H_5$—H. The ratio 5 H in aromatic ring : 2H in —$CH_2$— : 3 H in —$CH_3$ gives $C_6H_5C_2H_5$ as the compound.

5. Ethanal, $CH_3CHO$. The peak at $9.7-9.8\,\delta$ is due to RCHO. The peak at $2.1\,\delta$ is probably due to R—$COCH_3$. The ratio 1 H in —CHO : 3 H in $RCOCH_3$ gives $CH_3CHO$ as the formula.

6. Butan-2-one, $CH_3COCH_2CH_3$. The —$CH_3$ peak at 0.9–1.1 $\delta$ and the —$CH_2$— peak at 2.2–2.5 $\delta$ are identifiable. The peak at 2.1 $\delta$ can be due to $RCOCH_3$. A compound with a —$COCH_3$ group and a $C_2H_5$— group is $CH_3COCH_2CH_3$.

## Questions on Chapter 7

2. (a)  There is only one type of H atom. It must be $CH_3OCH_3$, methoxymethane. The NMR spectrum is different from that of $CH_3CH_2OH$ because in ethanol there are H atoms in three different environments, —OH, —$CH_2$— and —$CH_3$.
   (b)  Tetramethylsilane, $(CH_3)_4Si$, has all its H atoms in identical environments and gives a single NMR line.

3.  There are three types of H atom: two split into a quartet by 3 adjacent H atoms, could indicate —$CH_2CH_3$; three not split; three split into a triplet by 2 adjacent H atoms, e.g. —$CH_2CH_3$. The ester could be $CH_3CO_2CH_2CH_3$.

4.  The ratio of C : H indicates an aromatic compound. The high value of $\delta$ for one group of H atoms indicates deshielding by the benzene ring. There are H atoms in three environments. The ratio shown by the integration trace is 5:2:3.
The compound could be

**A**    $C_2H_5$    or an isomer of **B**    $CH_3$    $CH_3$

A has 5 H in the ring, 2 H in —$CH_2$— and 3 H in —$CH_3$.
B has 4 H in the ring and 6 H in 2(—$CH_3$).
The ratio 5:2:3 fits compound **A**, ethylbenzene.

5. (a)  For deshielding see §7.3.
   (b)  There are H atoms in three environments. Four are split into a quartet, indicating three adjacent H atoms;

therefore they must be two sets of 2 H, e.g. 2(—$CH_2$—). Six are split into a triplet, indicating two adjacent H atoms; therefore they must be two sets of 3 H, e.g. 2(—$CH_3$). The compound $C_4H_{11}N$ could contain NH, 2(—$CH_2$—) and 2(—$CH_3$). It could be $(CH_3CH_2)_2NH$, diethylamine.
   (c)  NMR can analyse at a distance, without consuming the sample, at body temperature and without damaging tissue.

6. (a)  Spectrum 1: Two singlets, one being 9× the other in intensity
Spectrum 2: A quartet and a triplet in the ratio 4:6
Spectrum 3: A singlet, a doublet and another doublet in the ratio 1:1:2 and a multiple peak.
   (b)  Spectrum 1 suggests 1 H and (—$CH_3$)$_3$.
Spectrum 2: A quartet suggests two —$CH_2$— groups adjacent to two —$CH_3$ groups. A triplet suggests splitting by a —$CH_2$— group. The ratio 4:6 means $2CH_2 : 2CH_3$.
Spectrum 3: The multiple peak shows a H atom adjacent to a C atom with groups containing many equivalent H atoms.
   (c)  1,    $CH_3$
        $CH_3 — C — OH$
                $CH_3$

   2,  $CH_3 — CH_2 — O — CH_2 — CH_3$

   3,    $CH_3$
        $CH_3 — C — CH_2 — OH$
                $H$

---

## CHAPTER 8: MASS SPECTROMETRY

### Checkpoint 8.5

1.  $n = 1.15/15.0 \times 100/1.1 = 7$, and the compound is $C_7H_6O_2$.
2.  The peak at 15 corresponds to $CH_3^+$. The peak at 94 is probably the molecular peak with a side-peak at 96 suggesting chlorine or bromine. Peaks at 79 and 81 of almost equal heights indicate bromine. $CH_3(15) + Br(79) = 94$, and the compound is bromomethane, $CH_3Br$.
3.  64 is probably the molecular peak with a side-peak of 66 due to the presence of $^{37}Cl$. There is a peak at 29 corresponding to the loss of $^{35}Cl$ from the molecular ion of $M = 64$. The peaks at 49 and 51 could correspond to the loss of $CH_3$ from $M = 64$ and from $M = 66$. The peaks at 29, 28 and 27 could be $C_2H_5^+$, $C_2H_4^+$ and $C_2H_3^+$. Combining $C_2H_5$ and Cl gives $C_2H_5Cl$, chloroethane.
4.  $220 = {}^{35}Cl_2C_6H_3O^+$, $222 = {}^{35}Cl^{37}ClC_6H_3O^+$, $175 = {}^{35}Cl_2C_6H_3OCH^+$, $177 = {}^{35}Cl^{37}ClC_6H_3OCH^+$, $162 = {}^{35}Cl_2C_6H_3O^+$, $164 = {}^{35}Cl^{37}ClC_6H_3O^+$,

### Checkpoint 8.7

1.  $C_7H_8NO$ (The other formulae give $M = 122.024\,201$ and $122.048\,010$.)
2.  $C_8H_6N_2O_4$ (The other formulae have $M = 194.1531$ and $194.0579$)

### Checkpoint 8.9

1. (a)  The IR spectrum shows a double peak of absorption at 1850–1750 $cm^{-1}$, the region where C=O absorbs, indicating two C=O groups. There are peaks at 3000 $cm^{-1}$ and 1200 $cm^{-1}$ which indicate C—H bonds and a peak at 1400 $cm^{-1}$ which is typical of C—C bonds. Two C=O groups often indicate an acid anhydride.
   (b)  The MS shows no molecular peak. The peak at 72 could arise from the loss of CO or $C_2H_4$ from the molecule. The base peak at 28 could be due to $C_2H_4^+$ or $CO^+$. The peak at 56 could arise from the loss of $CO_2$, leaving $C_3H_4O$. The peak at 44 could be due to $CO_2^+$ or $C_2H_4^+$.

(c) The NMR spectrum shows that the 4 H atoms are equivalent. This could mean 2 $CH_2$ groups in equivalent positions: $CH_2CH_2C_2O_3$.

Combining this information, the suggestion of an acid anhdride from (a) and the indication from (c) that the 4 H atoms are in 2 $CH_2$ groups gives the formula

for the compound, ethanedioic anhydride.

**Questions on Chapter 8**
1. (a) **A**, $HCO_2CH_2CH_2CH_3$; **B**, $HCO_2CH(CH_3)_2$; **C**, $CH_3CO_2C_2H_5$; **D**, $C_2H_5CO_2CH_3$
   (b) All give a molecular peak at 88.
   $HCO_2^+ = 45$, and this is absent, eliminating **A** and **B**.
   $CH_3CO_2^+ = 59$, and $C_2H_5 = 29$, and these are both present so **C** is a possibility.
   $C_2H_5CO_2^+ = 73$, and $CH_3 = 15$, and both these are absent, so **D** is unlikely.
   The ester is **C**, $CH_3CO_2C_2H_5$.
2. The peak at $m/e = 46$ is the molecular peak. The peak at $m/e = 47$ corresponds to $^{12}C^{13}CH_5OH$. The peak at $m/e = 45$ is due to $C_2H_5O^+$, 31 to $CH_2OH^+$, 29 to $C_2H_5^+$ and 27 to $C_2H_3^+$.
3. The peak at 136 is the molecular peak. The peak at 77 corresponds to $C_6H_5$. The peak at $105 = 136 - 31$ and could be due to the loss of $OCH_3$, leaving $C_7H_5O^+$. The peak at 51 could be $C_4H_3^+$. Combining the benzene ring and the ability to lose $OCH_3$ to form $C_6H_5CO^+$ indicates $C_6H_5CO_2CH_3$, methyl benzoate.
4. 58 = molecular peak, 57 = loss of one H = $C_3H_5O^+$, 29 could be $C_2H_5^+$ or $CHO^+$ 28 could be $C_2H_4^+$ or $CO^+$. Combining $C_2H_5$ and CO and H gives $C_2H_5CHO$, propanal.
5. 102 = molecular ion, peak at $87 = M - CH_3 = C_5H_{11}O^+$, peak at $73 = 87 - 14 = 87 - CH_2 = C_4H_9O^+$, peak at 57 could be $M - C_2H_5O = C_4H_9^+$, peak at 45 could be $C_2H_5O^+$, 29 = $C_2H_5^+$. Combining $C_2H_5O$ and $C_4H_9O$ gives $C_4H_9OC_2H_5$. The isomer is in fact 2-ethoxybutane,

$$CH_3CH_2CHCH_3$$
$$|$$
$$OC_2H_5$$

**Questions on Combined Techniques**
1. (a) 122 = molecular peak, 77 = $C_6H_5^+$, 105 corresponds to a loss of 17 which could be due to OH, making the ion $C_7H_5O^+$, which could be $C_6H_5CO^+$. The peak at 51 could be $C_4H_3^+$. Combining the groups $C_6H_5$ and $C_6H_5CO$ gives $C_6H_5CO_2H$, benzoic acid.
   (b) The absorption at 2500–3000 $cm^{-1}$ suggests an —OH group. This is supported by the peak at 1300 $cm^{-1}$. The absorption at 1700–1750 $cm^{-1}$ suggests a >C=O group. The absorption over 900–1100 $cm^{-1}$ could be due to the benzene ring or to a C—O bond. The peaks confirm $C_6H_5CO_2H$.
2. In the mass spectrum, 58 = molecular peak, 15 = $CH_3^+$, 43 = $M - CH_3 = C_2H_3O^+$ which could be $CH_3CO^+$, 42 = $C_2H_2O^+$. Combining $CH_3CO^+$ and $CH_3^+$ gives $CH_3COCH_3$. The IR spectrum shows

absorption at 1720 $cm^{-1}$, indicating a C=O group, at 2800–3000 $cm^{-1}$ and 1400–1500 $cm^{-1}$, indicating aliphatic C—H bonds.
3. The IR spectrum shows peaks at 2800–3000 $cm^{-1}$ and 1400–1500 $cm^{-1}$ which typify C—H bonds. There is an absorption at 700 $cm^{-1}$ which could be due to C—Cl. The mass spectrum shows a molecular peak at $m/e = 92$ and an $(M + 2)$ peak which is $\frac{1}{3}$ of the height of the $M$ peak and suggests Cl. Other peaks may be $CH_2^{35}Cl^+ = 49$, $CH_2^{37}Cl^+ = 51$, $C_3H_7^+ = 43$, $C_2H_5^+ = 29$, $CH_2CH_2CH_2^{35}Cl^+ = 77$, $CH_2CH_2CH_2^{37}Cl^+ = 79$. The compound is chlorobutane. (In fact it is 1-chlorobutane.)
4. The IR spectrum shows the absorption at 1700 $cm^{-1}$ due to C=O. A C—H absorption above 3000 $cm^{-1}$ can be seen and the absorption over 670–770 $cm^{-1}$ indicates an aromatic ring. So far R seems to be an aromatic aldehyde or ketone.
   The MS shows a molecular ion at $m/e = 120$. The peak at 105 corresponds to the loss of $CH_3$ from the molecule. The peak at 77 corresponds to $C_6H_5^+$, and that at 43 to $CH_3CO^+$. The peak at 51 is probably a fragment of the benzene ring, $C_4H_3^+$. Combining these fragments gives $C_6H_5COCH_3$, phenylethanone, which fits the inferences from the IR spectrum.
5. (a) (ii) 12.0
   (b) (i) 120 (ii) the presence of $^{13}C$
   (c) (i) $M/(M + 1) = 33/3.0 = 100/1.1n$; $n = 8$; the high ratio of C to H is typical of aromatic compounds.
   (ii) B, $C_6H_5^+$, C, $C_6H_5CO^+$
   (iii) When the bond $C_6H_5CO—CH_3$ breaks, $C_6H_5CO^+$ is formed in preference to $CH_3^+$ because the charge is delocalised in $C_6H_5CO^+$; ·$CH_3$ is formed as a radical, not an ion. When the bond $C_6H_5—COCH_3$ breaks, the $CH_3CO^+$ formed is resistant to fragmentation.
   (d) 2,4-Dinitrophenylhydrazine gives an orange precipitate.
6. IR shows an aromatic ring in the fingerprint region, O—H at 3500 $cm^{-1}$, C—H, aromatic at 3100 $cm^{-1}$ and C—H aliphatic at 2900 $cm^{-1}$. With the formula $C_7H_8O$, the compound could be $CH_3C_6H_4OH$.
   NMR shows three types of H atom: one H atom of one type, three H atoms of a second type and four H atoms of a third type. This fits in with the formula $CH_3C_6H_4OH$.
   The MS shows a molecular peak at 108, a peak due to $CH_3C_6H_4O^+$ a peak at 91 which could be due to $CH_3C_6H_5^+$, a peak at 76 which could be due to $C_6H_4^+$, and peaks which are due to fragments of the benzene ring such as $C_4H_5^+$ at 53, $C_4H_2^+$ at 50, $C_2H_3^+$ at 27.
7. (a) (i) See Figure 8.1A (ii) See §8.1
   (iii) $M = 46$, $M + 1 = 47$. There is a small % of $^{13}C$.
   (iv) 31 = $CH_2OH^+$, 29 = $C_2H_5^+$, 27 = $C_2H_3^+$
   (b) (i) See §7.5 (ii) See §7.5 (iii) See §7.3
8. (a) (i) NMR (ii) X-ray diffraction (iii) IR
   (iv) MS (v) visible–UV (vi) NMR. When X—H is substituted by D to form X—D, the compound does not have a NMR spectrum.
   (b) (i) Electrons are promoted to orbitals of higher energy.
   (ii) $A = \log(I_0/I)$
   (iii) $\varepsilon = A/cl$
   (iv) $[X/Y] = (2/1) \times \frac{1}{3} = \frac{2}{3}$

(v) Other substances in the water will absorb in the same region as **X** and **Y**. Use a double-beam spectrophotometer with the **X** + **Y** sample in one cell and the effluent containing no **X** and **Y** in the reference cell.

9. (*a*) (i)  See Figure 8.1A
   (ii)  in the ionisation chamber
   (iii)  The *M* peak and the (*M* + 2) peak are equal if the compound contains one Br per molecule.
   (*b*) (i)  Intermolecular hydrogen bonding gives —O—H ····· O— absorption at 3600 cm$^{-1}$. Structure **A** allows for H-bonding.
   (ii)  Triplet: protons which give rise to the signal are protons in —CH$_3$.
   Quartet: protons which give rise to the signal are protons in —CH$_2$—.
   (iii)  The proton in —OH in compound **A** will exchange with deuterium to form the group —OD which does not have an NMR peak. In **B** no such exchange is possible.

10. (*a*) (i)  **B** owing to conjugation   (ii)  molar absorption coefficient
   (*b*)  Intramolecular hydrogen bonding reduces the strength of the O—H bond and it vibrates at lower frequency.

(*c*) (i)  O—H proton. Exchange leads to the formation of O—D which does not give rise to an NMR spectrum.
   (ii)  Integration trace; **A**:**B** = 1:4
(*d*)  X-ray diffraction; measure the positions and intensities of the spots.

11. (*a*) (i)  Water absorbs IR and will therefore interfere with the spectrum of **C**.
   (ii)  The spectrum of pure **C** can be used for comparison.
   (iii)  See Figure 5.1B
   (iv)  a change in the dipole moment of the molecule
   (*b*) (i)  Bombardment of molecules with electrons; see Figures 8.1A, 8.6A.
   (ii)  Deflection in a magnetic field; see Figure 8.1A
   (iii)  The ratio (*M* + 1) peak/*M* peak [§ 8.5] gives the number of C atoms per molecule. The value of *M* found accurately at high resolution will distinguish; see § 8.7. Fragmentation patterns will distinguish; see § 8.4.
   (*c*) (i)  In (CH$_3$)$_4$Si, all the H atoms are in identical environments.
   (ii)  There are four spin combinations of CH$_3$ protons. They have an effect on the resonance of the CH$_2$ group; see § 7.6.

# Index